C0-AYH-262

Uncovering the Hidden Harvest

Valuation Methods for Woodland and Forest Resources

Edited by Bruce M Campbell and Martin K Luckert

Earthscan Publications Ltd,
London • Sterling, VA

First published in the UK and USA in 2002 by
Earthscan Publications Ltd

ISBN: 1 85383 809 8

Typesetting by PCS Mapping & DTP, Newcastle upon Tyne
Printed and bound in the UK by Thanet Press Ltd, Margate, Kent
Cover design by Yvonne Booth
Cover photo by Carol Colfer
Panda symbol © 1986 WWF
® WWF registered trademark owner

For a full list of publications please contact:
Earthscan Publications Ltd
120 Pentonville Road
London, N1 9JN, UK
Tel: +44 (0)20 7278 0433
Fax: +44 (0)20 7278 1142
Email: earthinfo@earthscan.co.uk
Web: **www.earthscan.co.uk**

22883 Quicksilver Drive, Sterling, VA 20166–2012, USA

Earthscan is an editorially independent subsidiary of Kogan Page Ltd and publishes in association
with WWF-UK and the International Institute for Environment and Development

A catalogue record for this book is available from the British Library

Library of Congress Cataloging-in-Publication Data

Uncovering the hidden harvest : valuation methods for woodland and forest resources / edited by
Bruce M. Campbell and Martin K. Luckert.
       p. cm. — (People and plants conservation manuals)
   Includes bibliographical references (p.   ).
   ISBN 1-85383-809-8 (pbk.)
   1. Forests and forestry—Economic aspects. 2. Forest products—Economic aspects. I. Campbell,
B. M. (Bruce M.) II. Luckert, Martin Karl, 1961— III. Title. IV. People and plants conservation
manuals (Earthscan Publications Ltd.)

SD393 .U48 2001
634.9'8–dc21

                                                                                    2001002397

This book is printed on elemental chlorine-free paper

Uncovering the Hidden Harvest

PEOPLE AND PLANTS CONSERVATION SERIES

Series Editor
Martin Walters

Series Originator
Alan Hamilton

People and Plants is a joint initiative of WWF,
the United Nations Educational, Scientific and Cultural Organization (UNESCO)
and the Royal Botanic Gardens, Kew
www.rbgkew.org.uk/peopleplants

Titles in the series

Applied Ethnobotany: People, Wild Plant Use and Conservation
Anthony B Cunningham

People, Plants and Protected Areas: A Guide to In Situ *Management* (reissue)
John Tuxill and Gary Paul Nabhan

Plant Invaders: The Threat to Natural Ecosystems (reissue)
Quentin C B Cronk and Janice L Fuller

*Uncovering the Hidden Harvest: Valuation Methods for Woodland and
Forest Resources*
Bruce M Campbell and Martin K Luckert (eds)

Forthcoming titles in the series

Biodiversity and Traditional Knowledge: Equitable Partnerships in Practice
Sarah A Laird (ed)

Ethnobotany: A Methods Manual 2nd edition
Gary J Martin

*Tapping the Green Market:
Management and Certification of Non-Timber Forest Products*
Patricia Shanley, Alan R Pierce, Sarah A Laird and Abraham Guillén (eds)

Contents

List of figures, tables and boxes

Figures

Tables

Boxes

List of contributors

Thomas Beckley, Faculty of Forestry and Environmental Management, University of New Brunswick

Brian Belcher, Center for International Forestry Research, Indonesia

Christopher Bennett, Department of Agricultural Economics, University of British Columbia, Canada

Ivan Bond, WWF Southern Africa Regional Programme Office, Zimbabwe

Peter C Boxall, Department of Rural Economy, University of Alberta, Canada

Neil Byron, Productivity Commission, Australia, and Center for International Forestry Research, Indonesia

Bruce M Campbell, Institute of Environmental Studies, University of Zimbabwe, and Center for International Forestry Research, Indonesia

William Cavendish, T H Huxley School, Imperial College of Science, Technology and Medicine, United Kingdom

Wil De Jong, Center for International Forestry Research, Indonesia

Sonya Dewi, Center for International Forestry Research, Indonesia

Antoine Eyebe, Center for International Forestry Research, Cameroon

Peter G H Frost, Institute of Environmental Studies, University of Zimbabwe

Bev Geach, Group for Environmental Monitoring (GEM), South Africa

Allison Goebel, Institute of Women's Studies, Queens University, Canada

Martin K Luckert, Department of Rural Economy, University of Alberta, Canada, and Center for International Forestry Research, Indonesia

Freddy R Mathabela, Bushbuckridge, South Africa

Ousseynou Ndoye, Center for International Forestry Research, Cameroon

Nontokozo Nemarundwe, Institute of Environmental Studies, University of Zimbabwe, and Center for International Forestry Research, Indonesia

Thiambi R Netshiluvhi, Environmentek, CSIR, South Africa

Calvin Phiri, AWARD, South Africa

Michael Richards, Forest Policy and Environment Programme, Overseas Development Institute, United Kingdom

Dede Rohadi, Center for International Forestry Research, Indonesia

Claudia Romero, University of Florida, United States

Manuel Ruiz Pérez, Center for International Forestry Research, Indonesia, and Department of Ecology, Autonomous University of Madrid, Spain

Charlie M Shackleton, Environmental Science Programme, Rhodes University, South Africa

Sheona E Shackleton, Environmental Science Programme, Rhodes University, South Africa

Bevyline Sithole, Centre for Applied Social Sciences, University of Zimbabwe

Terrence S Veeman, Department of Rural Economy, University of Alberta, Canada

Michele Veeman, Department of Rural Economy, University of Alberta, Canada

The People and Plants Initiative

This book is a contribution to People and Plants, an initiative designed to enhance the role of communities in efforts to conserve biodiversity and use plant resources in ways that can be sustained. The programme builds capacity through:

- providing opportunities to train young professionals, capable of working with communities to identify and resolve local issues relating to plant resources;
- supporting the development of institutions in civil society, such as non-governmental organizations and training establishments, which work with communities or train ethnobotanists;
- identifying and promoting wise practice approaches and methods in applied ethnobotany.

There are People and Plants field projects in Nepal and Pakistan, concerned variously with conservation of Himalayan medicinal plants, support for local health-care services based on traditional medicine, community involvement in forest management and the provision of alternative fuelwood resources. The purposes of a campaign in Kenya are to place the hardwood carving industry on a more sustainable basis, safeguard the livelihoods of the carvers and conserve forest biodiversity. Three themes have been selected for the identification of best practice approaches and methods: conservation and woodcarving; conservation of Himalayan medicinal plants; and community plant use and protected areas. We are also exploring ways to develop curricula in applied ethnobotany.

People and Plants maintains a website with much information on the programme, including details of its various publications, of which some are downloadable. Visit the website at: www.rbgkew.org.uk/peopleplants.

People and Plants is a partnership of WWF and UNESCO. The Royal Botanic Gardens, Kew, is an Associate, especially concerned with provision of information.

Alan Hamilton
Head, International Plants Conservation Unit
WWF-UK

PEOPLE AND PLANTS WEBSITE:

www.rbgkew.org.uk/peopleplants

People and Plants partners

WWF

WWF (formerly the World Wide Fund For Nature), founded in 1961, is the world's largest private nature conservation organization. It consists of 29 national organizations and associates, and works in more than 100 countries. The coordinating headquarters are in Gland, Switzerland. The WWF mission is to conserve biodiversity, to ensure that the use of renewable natural resources is sustainable and to promote actions to reduce pollution and wasteful consumption.

UNESCO

The United Nations Educational, Scientific and Cultural Organization (UNESCO) is the only UN agency with a mandate spanning the fields of science (including social sciences), education, culture and communication. UNESCO has over 40 years of experience in testing interdisciplinary approaches to solving environmental and development problems in programmes such as that on Man and the Biosphere (MAB). An international network of biosphere reserves provides sites for conservation of biological diversity, long-term ecological research and testing and demonstrating approaches to the sustainable use of natural resources.

ROYAL BOTANIC GARDENS, KEW

The Royal Botanic Gardens (RBG), Kew, has 150 professional staff and associated researchers and works with partners in over 42 countries. Research focuses on taxonomy, preparation of floras, economic botany, plant biochemistry and many other specialized fields. The Royal Botanic Gardens has one of the largest herbaria in the world and an excellent botanic library.

The African component of the People and Plants Initiative is supported financially by the Darwin Initiative, the National Lottery Charities Board and the Department for International Development (DFID) in the UK, and by the Norwegian Funds in Trust.

DISCLAIMER

Acknowledgements

This work was undertaken in the context of ongoing cooperation between the Universities of Alberta and Zimbabwe, cooperation that was initially funded by the International Development Research Centre (IDRC) through the Value of Trees project, and that is presently funded by the Canadian International Development Agency (CIDA) through the Agroforestry: Southern Africa project.

The production of this book has been funded by the People and Plants Programme, a joint endeavour of WWF, UNESCO and the Royal Botanic Gardens, Kew.

We acknowledge the assistance of the following reviewers of chapters: Arild Angelsen (Agricultural University of Norway), Will Cavendish (Imperial College), Kevin Chen (University of Alberta), Tony Cunningham (WWF), Cathy Dzerefos (University of Witwatersrand), Wil de Jong (Center for International Forestry Research), Isla Grundy (University of Stellenbosch), Naomi Krogman (University of Alberta), Jack Putz (University of Florida and Center for International Forestry Research), Bev Sithole (University of Zimbabwe), Terry Veeman (University of Alberta), Knut Veisten (Agricultural University of Norway), Ed Witkowski (University of Witwatersrand) and Sven Wunder (Center for International Forestry Research).

Thanks for funding some of the field research contributing to the book are due to the Forestry Research Programme of the UK Department for International Development, the German Federal Ministry of Economic Cooperation and Development (BMZ), the International Development Research Centre (IDRC), the Canadian International Development Agency (CIDA), that part of the People and Plants Programme funded through WWF, and the Center for International Forestry Research (CIFOR).

We thank Widya Prajanthi Manoppo of the Center for International Forestry Research for preparing the figures for publication.

Chapter 1

Towards understanding the role of forests[1] in rural livelihoods

Bruce M Campbell and Martin K Luckert

Introduction: the partially hidden harvest

Over the last decade, increasing attention has been paid to socio-economic aspects of rural development and to understanding the social processes in rural systems in developing countries. Whereas previously, technical solutions were frequently seen as the key to successful development, there has been growing recognition of the need to understand social dimensions of development to ensure that interventions lead to improvements in the livelihoods of households.

One key area where a greater understanding of social dimensions has been sought is with regard to the values of goods and services derived from indigenous plants and animals (Table 1.1). The International Institute for Environment and Development (IIED) coined the phrase 'the hidden harvest' (eg IIED, 1995; Hot Springs Working Group, 1995) to draw attention to the importance of indigenous species in influencing the livelihoods of households. Some of the initial efforts to uncover the hidden harvest involved descriptive studies of which indigenous species were used by households (see Cunningham, 1997, for a review of much

Table 1.1 *A classification of forest values*

Type of value	Sub-types of values	Examples
Direct use	Consumptive	*Commercial goods*, eg timber, fruit, animals, rattan, medicine, firewood for sale, charcoal *Non-market goods*, eg firewood for subsistence use (Figure 1.1), subsistence foods (Figure 1.2)
	Non-consumptive	Recreation, eg ecotourism; forest research and education, shade
Indirect use		Habitat protection, watershed and soil protection, carbon sequestration
Optional use		Maintaining options by avoiding irreversible damage to soils, water sources; maintaining stocks for future use
Non-use or passive use	Existence use	Knowledge of a sacred site that no one is allowed to visit
	Bequest to future generations	Passing on natural resources for the support of future generations

Photograph by Carol Colfer

Figure 1.1 *A woman in Appoisso, Côte d'Ivoire, carrying wood back to her home for cooking and heating. For this subsistence use of wood, the questions that this book attempts to answer include: how can the amount used per year be quantified; what is the appropriate value of the good (per headload and per kg) and how can the labour be valued (given that other activities were undertaken during the course of collection); and how can the labour be priced?*

Photograph by Manuel Ruiz Pérez

Figure 1.2 *The dry savannas of southern Africa are a source of numerous forest products. Here three young boys return with their harvest of baobab fruits, collected while herding cattle, Hot Springs, Zimbabwe*

of this literature; see also Box 2.1 in Tuxill and Nabhan, 2001, p13). The number, diversity and intensity of use presented in these studies generally gave the impression that these types of resources were playing key roles in the livelihoods of local peoples. Although these descriptions have been crucial first steps to understanding the role of forests in rural livelihoods, the significance of the harvest has still remained partially hidden. To make direct comparisons between the values of indigenous species and those of agricultural species, monetary values were sought.

Historically, development economists have been actively involved in attempting to create a macro-economic framework suitable for development. However, more recently, applied micro-economists have begun using their tools in developing country settings to better understand the

connection between resources and livelihoods of local peoples. One set of tools that these economists have brought includes valuation techniques. A number of different methods, such as using related prices (see Chapters 2 and 4), contingent valuation and travel cost techniques (see Chapter 4), have been used, depending upon the specific circumstances of the valuation questions under consideration. These methods have also been supplemented by inquiries into market structures (see Chapter 3), cost-benefit analyses (see Chapter 5), and other methods such as those included in participatory rural appraisal (PRA) (see Chapter 6).

Estimating values of forests and deriving the household incomes from these resources is a key step towards understanding the role of forests in rural livelihoods. The desire to evaluate the

3

BOX 1.1 THE VALUE OF WOODLANDS IN NAMIBIA

In a widely circulated booklet from the Directorate of Forestry (1996) in Namibia, the values of trees and woodlands in Namibia are presented. The data show that these resources have many different values, and that in total they are worth some N$105.8 million per year* (Figure 1.3). Of particular importance are wood values (for fence poles, fuel and construction), but tourism is also important. In the same booklet the government allocation to the Directorate of Forestry is given as N$3 million in 1996. The preface to the document is by the president of the country.

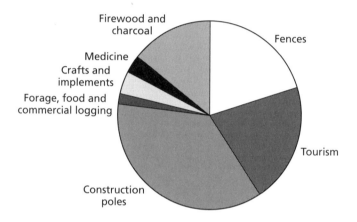

Figure 1.3 *The distribution of the total economic value of forest resources in Namibia*

The document represents part of the strategy by the directorate to raise understanding about forest values and to build the profile of the forest sector. The importance of forest products other than commercial timber is amply demonstrated. The valuation exercise has done much to raise the profile of the Directorate of Forestry and to indicate to a wider audience the importance of trees and woodlands to the national economy.

* In 1996 US$1 = N$4.36.

contribution of indigenous species to household livelihoods is largely based on a desire to produce policy-relevant research. A number of potential uses for value estimates of indigenous resources are perceived. Firstly, there are land-use questions where, for example, centrally planned attempts have sought to improve livelihoods by replacing indigenous species with agricultural species. To weigh the trade-offs associated with forgone indigenous resources, value estimates are necessary (see Box 5.1). Secondly, there are

development projects involving indigenous species where cost-benefit analyses need to be undertaken. Values of indigenous species are crucial in attempting to get as broad a measure of social welfare as possible. Thirdly, value estimates are used to justify and argue for a change in focus, towards greater consideration of indigenous resources in rural development. We present two examples of this, one using data for the national level for Namibia (Box 1.1) and the other using data from three case study areas in South Africa (Box

BOX 1.2 DIRECT-USE VALUE OF WOODLAND RESOURCES IN THREE RURAL SETTLEMENTS, SOUTH AFRICA

Charlie M Shackleton, Sheona E Shackleton,
Thiambi R Netshiluvhi and Bev Geach

Woodlands cover one third of South Africa and are home to almost a quarter of its population, mostly in the densely populated communal areas of the former homelands. As this box will demonstrate, woodlands can make a significant contribution to local and national production and economic growth. Many of the products collected from woodlands are consumed at home or traded only locally, and very few products find their way into formal markets. Consequently, there is little appreciation in government and planning fora of their importance in rural livelihoods and their contribution to the informal economy.

Case studies of resource use in three widely separated rural settlements of differing economic and development status were recently conducted (Shackleton et al, 1999a; Shackleton and Shackleton, 2000). On a relative scale Mogano (Northern Province) was the most developed settlement, and of the three it had the highest employment levels. There was trade in most woodland products to a greater or lesser extent. In comparison, KwaJobe (in KwaZulu-Natal) was a relatively isolated settlement far from local economic centres. Ha-Gondo (Northern Province) was intermediate between these two, sharing characteristics of them both, but more akin to KwaJobe.

Within each of these villages three complementary techniques were used to gather information to allow determination of local direct-use values: semi-structured interviews, key-informant interviews with specialist resource users (such as traditional healers, carvers, vendors), and PRA exercises in general community meetings. The interviews focused on which resources were used, frequency of use, the amounts used per person or per household, local prices, whether or not respondents ever purchase forest products, harvesting areas and changes in resource supply.

Households in each of the settlements used a wide range of products, although a greater array of products was used in the least developed settlement, KwaJobe (Table 1.2). Each village was characterized by specific product-use patterns. For example, use of indigenous wood for housing was high in KwaJobe and Ha-Gondo, but absent in Mogano. Mushrooms, fish and ilala palms were widely used at KwaJobe but not the other two settlements. Most households at Ha-Gondo consumed edible insects, but only about 50 per cent of households did so at the other two settlements. Averaged across the three settlements, the top five most widely used biotic products were edible herbs (97.3 per cent), wooden utensils (96.9 per cent), fuelwood (91.5 per cent), wild fruits (82.7 per cent) and hand brushes made from grass or ilala palm leaves (78.7 per cent).

Table 1.2 *Range of woodland products used in three rural settlements in South Africa*

	Mogano	Ha-Gondo	KwaJobe
Number of products used	20	18	27
Number of products with ≥ 90% of households using	4	6	10
Simpson's diversity index (1/D)*	11.5	12.0	19.4
Number of species used (excluding medicinal species)	118	208	170

* $1/D = 1/\Sigma p_i^2$, where p_i is the proportion of product 'i' in the sample.

There also was considerable variation in the mean amounts used per person of particular products across the three settlements. For example, at KwaJobe mean fuelwood use was 462 kilograms (kg) per person per year. Corresponding figures for Ha-Gondo and Mogano were 1038kg and 885kg respectively. The number of axe-handles crafted and used from indigenous wood was 0.48 per household in KwaJobe, 0.10 in Ha-Gondo and 0.02 in Mogano. Similarly, the local unit price of different products differed between the villages by as much as an order of magnitude for some (Table 1.3).

Table 1.3 *Examples of mean unit prices (South African rand)* * for selected products in three settlements in South Africa*

Products	Units	Mogano	Ha-Gondo	KwaJobe
Fuelwood	per kg	0.45	0.22	(no trade)
Wild fruits	per kg	3.76	6.77	(no trade)
Edible herbs	per kg	34.48	33.94	2.65
Edible insects	per mug	2.94	2.82	2.00
Grass hand brushes	per item	5.61	3.41	4.00
Hoe handle	per item	17.24	6.80	5.30

* An exchange rate of US$1.00 = R5.70 is applicable.

Given the differences in the range of products used, amounts per capita and local prices, the total direct-use value differed considerably between the three villages: R1654 per person per year in Mogano, with corresponding figures of R671 and R375 in Ha-Gondo and KwaJobe, respectively. The top five products contributed 96.0 per cent of the total value in Mogano, 90.2 per cent of the total value in Ha-Gondo and 73.6 per cent of the total value in KwaJobe. Three products were repeated in the top five most valuable for each of the three villages. These were fuelwood, edible herbs and edible fruits. Differences in the amount of edible herbs consumed and their unit price accounted for a large proportion of the differences in the total value between the villages. Annual per capita consumption was 29.7kg, 20.5kg and 11.5kg for Mogano, Ha-Gondo and KwaJobe, and the unit price at KwaJobe was less than 10 per cent of that at the other two villages.

In each of the three case studies there was noticeable trade in several products, but with little commonality between the three sites. Overall, residents of Mogano purchased 13 different products and sold 9. Corresponding figures for Ha-Gondo were 10 and 5, respectively, and at KwaJobe were 11 and 14, respectively. Thus, residents at KwaJobe sold a greater range of products than they purchased, which was not the case at the other two villages. Additionally, a greater proportion of households sold some products. At Mogano and Ha-Gondo, less than 6 per cent of households were involved in selling any specific woodland product, whereas at KwaJobe eight products were traded by more than 6 per cent of households. The value of trade to specific households was not adequately captured within the research approach. In all three villages, specific householders were encountered (during the household or group work) who stressed the importance of being able to sell woodland products as a means of generating income. In some instances it was only one or two products, but some householders would sell whatever products for which they perceived there to be a demand. Usually, but not always, these households had no other means of cash income other than the sale of agricultural products.

Results from these three case studies corroborate other work in southern Africa demonstrating the extensive use of woodland resources by rural households, and their importance in financial terms to rural livelihoods. Collection of these resources not only represents a saving of scarce cash resources but also an opportunity for income generation either on a regular basis or during times of need. While total direct use values do not necessarily decrease with the level of development of rural villages, the composition of that value changes. In better-developed villages, the proportion of households purchasing woodland products is greater and the unit prices are higher. The significant direct-use value indicates that greater attention is required in supporting initiatives aimed at sustainable use of woodlands. In South Africa, provincial government provides extension support to agriculturalists in the crop and livestock sector. Similar recognition is required of the value of natural resource harvesting as another component of rural livelihood strategies.

1.2). In the former, the aim of the valuation was to support the contention that the forestry sector must be taken seriously in Namibia. The South African case study material was used to uphold the assertion that woodlands can make a significant contribution to local and national production and economic growth in South Africa. Fourthly, values of resource use can provide important insights into people's resource management options and livelihood strategies (see Chapter 2).

Characteristics of rural households and woodland products

To understand the role of forest products in households we need to understand the nature of rural livelihoods and the characteristics of forest products. Rural households typically have a wide livelihood portfolio, encompassing a range of activities (Cavendish, 2001). It is not uncommon for a household to be involved in livestock-raising; growing a diversity of crops; collecting forest products for subsistence needs and sales; being involved in a variety of reciprocal transactions with fellow community members; having one family member in off-farm employment who remits money back to the household and having another member involved in some small-scale industry (eg brick-burning, carpentry, craft production, beer-brewing). Rural households generally face low availability of capital, are prone to risks and have little formal education (Cavendish, 2001; Padoch, 1992; Browder, 1992). Risks can be in the form of extreme weather conditions which can decimate crop and livestock production, jolts to the household economy brought on by illness, and rapid changes in the external economic climate (eg changes in prices, reduced numbers of tourists). While indigenous knowledge about the environment and resource use may be high in rural areas, skill levels for dealing with external markets are generally low.

What are some of the characteristics of forest products? Many forest products are common pool resources, with some showing very little exclusivity (Cavendish, 2001). Many of them can be brought into a marketing chain with minimum capital investment. In the face of risk, forest products are often a source of sustenance or can be used to raise cash in the case of emergencies. Most forest products do not require high skill levels to bring them into production.

There is thus a strong match between the characteristics of the rural poor and the characteristics of forest products (Cavendish, 2001). It can be predicted that it will be the poorest of the poor who will rely mostly on forest products, and this has been demonstrated in a number of studies (eg Arnold and Ruiz Pérez, 1998). Where forest resources need greater capital investment, then it is the wealthier households who are able to exploit them. Thus, for example, Cavendish (1997) has shown that poorer households more generally produce small carpentry items, while more wealthy households make larger items. Small animals (eg insects, birds) are relatively more important to the poor while larger animals (eg antelope) are more important to the wealthier households. There are often strong income elasticities for forest products.[2] In areas where rapid changes in household welfare

occur, it can be expected that there will be rapid changes in forest use patterns (in the case of Box 1.2, total value remains the same in villages with different socio-economic status, but the composition of that value changes dramatically across villages).

Cavendish (2001) states that from the perspective of rural households, there is nothing special or unique about many forest products (he was mainly writing about consumptive values). In fact, rural households use most forest products for prosaic reasons. That is to say, most forest products are goods which households utilize according to the economic nature of the product in question. Each forest product has particular physical, technical and economic characteristics that may or may not make households desire to use them. Decisions to use will be made in terms of the labour investment required, the labour availability in the household, the quality of the good, the quality and accessibility to substitutes, and the rules and regulations governing use (including cultural and social norms).

Linking to broader issues

While this book is focused on households, it is always necessary to see the household within the broader framework of, for example, national policies, of a specific macro-economic framework, of international tourist markets and of a global climate (Campbell et al, 2000c). Thus household behaviour is partly driven by macro-economic policies, and by legal frameworks which define the institutions operative in rural areas.

Some of the important linkages that need to be understood are described briefly below. Remittances have been

shown to be an important component of the incomes of rural households, particularly in Africa (eg Cavendish, 1997; Cliffe, 1992; Scoones et al, 1996). The levels of remittances are largely dictated by macro-economic factors. We need to understand how remittances are used and what local investments they are driving. If remittance flows decrease, what are the impacts for forest product use? Markets in many forest products are driven by external factors such as international prices, macro-economic conditions affecting tourism levels, unemployment levels and

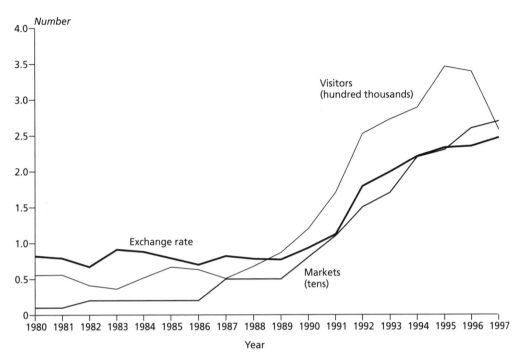

Note: The collapse in the Zimbabwean dollar caused a massive influx of South African visitors to Zimbabwe, and resulted in a major increase in woodcraft trade.
Source: Braedt and Standa-Gunda, 2000

Figure 1.4 *Number of visitors (tourists and people on business) entering southern Zimbabwe at Beitbridge; the change in the value of the Zimbabwean dollar relative to the South African rand; and the number of roadside craft markets on the Beitbridge–Masvingo Road, 1980–1997*

AIDS (eg Campbell et al, 2000c). To understand the role that forest product sales can play in the livelihood portfolio, we need to understand the international and national markets for such products and how they are likely to change.

For many forest products, tourism is important. The volatility of the tourism industry, linked to socio-political factors, needs to be understood. An example of the linkage between the external environment and rural livelihood strategies is demonstrated in Figure 1.4. With the onset of the structural adjustment programme in Zimbabwe, the Zimbabwe dollar was heavily devalued, one of the consequences being the rise in tourists from neighbouring South Africa. In some rural areas close to tourist routes this led to a massive rise in woodcraft sales (Figure 1.5). Policies set the scene for the institutional arrangements that are operative – they set the framework for the structure of property rights, the level of control in central versus local administrative structures, and so on. In this book we do not concentrate on these broader linkages; however, valuation is often pursued in order to understand the linkages.

Photograph by Tony Cunningham

Figure 1.5 *Carved wooden animals (hippos) made from* Afzelia quanzensis, *Zimbabwe*

Purpose and scope of the book

This book is directed to non-economists working in the context of developing countries. The purpose is to provide an overview of methods that may be used to assess the economic importance of forests to household livelihoods. The methods are presented with a number of examples of their use, most of them drawn from developing countries.

The importance of an interdisciplinary approach is now well recognized, and non-economists are increasingly working in teams that include economists. Moreover, many non-economists are being asked to plan and execute projects that include an economic element. However, the intention is not to provide step-by-step guidelines (see the following section on 'Importance of remaining observant and critical'). This book will provide the reader with the tools to understand the different approaches and methods and make more informed decisions as to which methods may be applicable.

The book presents eight chapters, which cover the following material:

- Chapter 1 gives us an indication of the importance of valuation, and the value of forests in rural livelihoods. It also cautions against the limitations of valuation.
- Chapter 2 shows how values can be derived for a full spectrum of environmental goods used by households, if prices are available for the goods.[3] The chapter is also useful for those undertaking household questionnaires, and demonstrates how environmental resources can be included within complete household accounts.
- Chapter 3 examines markets for goods from forests. The chapter gives an overview of some basic concepts of markets and market operations and describes some general features of rural markets. This is followed by a discussion of economic perspectives of market performance and an overview of market research tools.
- Chapter 4 turns to goods and services that are better tackled through the use of non-market valuation methods. It sets the stage for why non-market valuation may be important in the developing world, provides readers with some background economic theory that underlies non-market valuation, describes welfare measures, and introduces the range of non-market valuation methods.
- Chapter 5 describes some alternative ways that market and non-market values can be used within economic decision-making frameworks, focusing on variants of cost-benefit analysis, to provide understanding of rural livelihood systems and information relevant to policy-making.
- Chapter 6 examines the potential and limitations of tools drawn from participatory rural appraisal (PRA) to value forest resources in supplementing more traditional economic approaches. The authors conclude that PRA can be excellent for setting the context, but should not generally be used to derive quantitative data or monetary values.
- Chapter 7 examines the current state of interdisciplinary research and offers insights and suggestions as to how genuine synthesis among disciplines can be achieved. The underlying thesis is that an interdisciplinary approach is both essential and inevitable if research on the role of forests in rural

livelihoods is to be properly responsive to the needs of policy-makers and others.

- Chapter 8 concludes by indicating that resource values are only a small part of our full understanding of resources and livelihoods. A greater understanding may be derived from modelling household behaviour. It also suggests that we need to expand our conceptual and methodological approaches to livelihoods, which takes us beyond economics and into integrative methods, such as systems analysis.

Importance of remaining observant and critical

There is a danger that the numbers derived by valuation become ends in themselves. 'Eagerness to find monetary values may, by generating unrealistic numbers, eventually undermine the credibility of forest valuation in general' (Kengen 1997) (see Box 1.3). Furthermore, a valuation exercise is only one part of a much broader understanding that is required. While writing a book on valuation methods is one step towards promoting insight of resource values and livelihoods, there is the danger that we promote a rather simplistic notion of how to understand rural livelihood strategies. This topic is expanded in other chapters in the book, which stress the need to understand the complexity inherent in rural livelihood strategies, to fully understand the context of particular case studies and to use valuation as a first step towards understanding decision-making within rural households (see Chapters 6, 7 and 8).

The danger of non-economists attempting to undertake full-blown resource valuation exercises cannot be overstated (see last paragraph in Box 1.3). In addition, valuation methods are often complex or underdeveloped, and thus the nuances may be glossed over or there may be misapplication of concepts. There are numerous examples of inappropriate valuation. We have highlighted some of these in this book so that, hopefully, mistakes are not repeated (see, for example, Box 1.4, and the discussion in Chapter 6 on Box 6.5). Valuation requires strong disciplinary skills in economics. However, valuation exercises that do not include ecologists and sociologists, or researchers from similar disciplines, are likely to exclude important phenomena and processes. For instance, without an understanding of the complexities of fruiting in tropical trees, valuation could be based on inappropriate production data (see Box 7.6).

The authors of the chapters in this book are drawn from the disciplines of resource economics, ecology, microeconomics, agricultural economics and sociology, and most have been interacting on research projects in Zimbabwe for the last decade. Valuation has provided one of the foci for our interaction. We have found the interaction stimulating – and hope that this book captures some of that inspiration.

Box 1.3 Valuing resource valuation

*Neil Byron and Christopher Bennett**

'Not everything that can be counted counts,
and not everything that counts can be counted'
Albert Einstein (1879–1955)

When the history of ecosystem destruction and degradation in the 20th century is written, the blame for the inexorable loss of vast tracts of nature may be ascribed, in part, to inadequate understanding of the true value of these natural resources. Total economic valuation has been proposed as a means for making informed decisions (Carson, 1998; Kengen, 1997). We argue that there are persistent indications that resource valuation theory and practice suffer from a number of drawbacks.

Methodological difficulties
The difficulty of applying many of the methods has been noted and questioned (see, for example, the different chapters in this book; McFadden, 1994; Blamey et al, 1995; Boxall et al, 1996; Eberle and Gregory, 1991; Diamond and Houseman, 1994).

Shortcomings of resource valuation theory
Most valuation studies focus on environmental systems and impacts, as well as technical relationships. The complexities of social systems and the valuation of social goods and services are seldom considered fully. A one-dimensional price mechanism cannot reflect the divergent interests of individuals, groups and classes.

High cost of undertaking full resource valuation
The human and financial resource investments required for the design and implementation of comprehensive resource valuation are high. Data and knowledge about interactions between the economy and the environment are frequently inadequate. There are shortcomings in identification of multiple products and the impacts of their exploitation. Scientific knowledge about the production of, and demand for, environmental products and services is incomplete, often resulting in dubious estimates about highly uncertain and complex phenomena. Meanwhile, alternative approaches, which could have equal or greater potential for informing decision-making and resolving conflicts, may be undervalued or overlooked, such as proper and credible tenurial assurances for different kinds of forest managers, the need for institutional strengthening or the role of land-clearing subsidies in deforestation.

Small role in actual resource allocation decisions
It is probably fair to say that to date most resource allocation decisions in most countries have not been made on the basis of resource valuation, even in industrialized economies where the methodology is commonly applied and there are many practitioners. There are many determinants of policy change other than arguments generated by the results of resource valuation (eg political expediency, equity considerations, international pressure and conventions).

Monetary values are not everything

Even if all or most natural resource values can be identified and captured, some sustainable resource allocation decisions can and should be made without sole reference to economic conditions whether or not in terms of explicit monetary value. If commercial, large-scale logging in state forests is found to generate more public and export revenue, should this land use therefore be preferred by decision-makers to more traditional local uses because of the (allegedly) higher resource value of (assumed) sustainable commercial logging? The attempt to express most, if not all, values in monetary terms for ease of comparison is appealing but can lead to gross miscalculations and, ultimately, inappropriate decisions about resource allocation.

Poorly executed valuation studies

There are far too many 'bad' or poorly executed valuation studies, undertaken by people who think they have understood the fairly simple logic of the process but who have failed to grasp that the 'devil is in the detail' – that it is a very easy analysis to do poorly, but extraordinarily difficult to do well (Hanemann, 1994; Diamond and Hausman, 1994). This point is particularly relevant in the developing country setting where there are pressures to undertake more valuation studies, and yet there is little reliable data and few resource economists. A key question surrounding forest resource valuation is whether it will, like money, become a good servant or a bad master.

* For the full paper, see Byron and Bennett (1998).

Box 1.4 Valuing the rainforest – the hazards of valuation

William Cavendish

In one of the most cited papers in the literature on valuation of forest resources, Peters et al (1989), attempted to value rainforest goods. To do this they undertook a systematic botanical inventory of 1 hectare (ha) of forest 30 kilometres (km) from Iquitos, Peru. Within the hectare they found 72 species (with 350 individual plants) yielding products with a market value in Iquitos, comprising 11 fruit-bearing species, 60 timber species and 1 latex-producing species, *Hevea guianensis*. Valuation of these products proceeded as follows. Annual production rates of fruits were either measured directly or estimated through interviews with local people, and fruit prices were derived from monthly surveys of the Iquitos market (Box 7.6 notes the problems of estimating ecological functions). Yields of rubber from *Hevea* were taken from the literature and rubber prices were obtained from the agrarian bank office. The merchantable volume of timber was calculated using published regression equations and the mill price of each timber species acquired from local sawmills. Collection times for fruit and latex were estimated from interviews and direct observation, and the minimum daily wage in Peru was used to convert this to a monetary value. Transport costs for fruit and latex were estimated, based on previous studies, to be 30 per cent of total market value and 40 per cent for timber. Finally, sustainable fruit, latex and logging production levels were calculated. For logging, this implied a maximum sustainable harvest of 30 cubic metres per hectare (m^3ha^{-1}) every 20 years, so as to avoid damage to fruit and latex species. For fruit and latex, 25 per cent of annual production is left in the forest for regeneration.

With these assumptions, and using a 5 per cent discount rate, the authors calculated[*] that such sustainable forest production would generate US$341ha^{-1}yr^{-1}. Of this, fully US$300 is derived from fruit production and US$16.5 from latex, so fruits and latex together comprise more than 90 per cent of the total sustainable economic rent available from the forest. The authors then compare this with the returns to plantations (US$159ha^{-1}yr^{-1}) and cattle ranching (US$148ha^{-1}yr^{-1}), and conclude that the results clearly indicate the importance of non-timber forest products (NTFPs). They argue that 'without question, the sustainable exploitation of non-wood forest resources represents the most immediate and profitable method for integrating the use and conservation of Amazonian forests' and that Amazonian conservation could rest on comparative economics if the true value of non-wood resources was recognized. This study spawned several similar studies (eg Grimes et al, 1994; Balick and Mendelsohn, 1992) that came to comparable conclusions.

However, these conclusions rest completely on the assumptions underpinning the economic valuations undertaken. Unfortunately, there are significant problems here. Specifically, it is wrong to conduct an inventory of a hectare of forested land and assume that the entire inventory (or the share usable as sustainable production) has economic value. For example, if the entire forest supply of the many hectares around Iquitos came on the market, prices would decrease quite dramatically, so that current prices are therefore quite inappropriate for the task of valuing potential production. The problem here lies in the use of potential data (based on an inventory) rather than actual data (based on observed behaviour). If potential data are used, then great care has to be made in applying current price information, since these prices are based on existing market transactions rather than the potential levels of production and consumption being assumed. One way around this is to estimate how prices might respond to changing use levels: as an example, variable price and cost information was used for estimating firewood values in Campbell et al (1997b). But using valuations derived from observed rather than assumed behaviour (see Chapter 2) means that this problem can be avoided.

The point is that economic value really derives from what people are actually doing. For example, if the fruits in the Peters et al study were not being collected because there was no demand for them, or because the high costs of collection, storage and transportation made the supply of fruits unprofitable, then these fruits do not have a positive economic value at current prices, incomes, collection costs, etc.

Surveys based on actual usage report much lower values of NTFPs in the Amazon. Godoy et al (1993) cite studies estimating NTFP values at about US$40ha^{-1}yr^{-1} while Lampietti and Dixon's (1995) comprehensive review of case studies suggested an average NTFP value of US$50ha^{-1}yr^{-1}. In complete opposition to Peters et al, these studies would suggest NTFP-based forest conservation is far less economically valuable than plantations or cattle ranching. Indeed most recent studies concur that the problem with tropical forests is that NTFPs are nowhere near valuable enough, in terms of current use values, to hold their own against other land uses (eg Southgate, 1998; Anderson, 1992; see also Box 5.1). So the assumptions used in valuation studies can make huge differences to valuation estimates, and hence to policy conclusions as well.

* In the original article net present value (NPV) was used as the economic criterion to compare scenarios, but in order to allow comparison to other studies this has been converted to $/ha/annum. Sustainable timber yield is dealt with by smoothing out the income streams as if equal portions were being received each year. This is why a discount rate is needed (see 'Selecting among decision criteria' and Choosing a discount rate' in Chapter 5 for discussion on NPV and discount rates, respectively). Peters et al (1989) select a 5 per cent discount rate; it is likely that this rate represents a social discount rate rather than a private rate (see 'Choosing a discount rate' in Chapter 5 for a discussion on which rate is more appropriate under which circumstances). Choice of a higher discount rate would lessen the contribution of timber.

Notes

1   We have used the term 'forest goods and services' as shorthand for the goods and services provided for by savannas, woodlands and forests, and indigenous trees left in household and cropping areas; the products include wood, fruit, mushrooms, insects, mice, medicinal plants, etc.
2   That is, when income rises, the demand for many forest products changes sharply.
3   Here the term 'environmental' goods is used synonymously with 'goods from forests' except that for the former the author includes use of reeds from riverbeds, use of soil from termite mounds and other uses, which may not be strictly speaking from 'forests'.

Chapter 2

Quantitative methods for estimating the economic value of resource use to rural households

William Cavendish

Introduction

Why attempt to estimate the economic value of environmental resource use to rural households? There are three main reasons for collecting quantitative data in this area. Firstly, rural households often use environmental resources extensively (and these are largely plant resources), so that such resources can be critical to the well-being of these households, especially during droughts and other times of crises (Cavendish, 1999a; Martin, 1995; Wollenberg and Nawir, 1998). However, there have been few systematic attempts to measure the value of natural resources to rural households, and there are thus many outstanding questions about the workings of the rural economy. For example, who depends upon plant resources? Are they rich or poor? Male of female? Young or old? What might happen to the rural economy if plant resources disappeared? What might happen to the poor? As these questions suggest, there are major uncertainties about how plant resource use interacts with the incomes and welfare of rural dwellers.

Secondly, given the extent of environmental resource use, many forms of rural environmental degradation – such as deforestation, biodiversity loss, soil erosion and watershed degradation – will be driven by households' choices concerning environmental resource use. Rural households choose to degrade environments or, alternatively, to invest in environmental protection through planting, managing or changing use practices, because it is in their economic interests to do so. Their actions partly reflect the impact that national policies, tenure conditions and institutions have on the incentives facing individuals and households. But their actions are also strongly related to the economic conditions of individuals and households at the local level – for example, their poverty, their economic opportunities, their educational status and so on. To understand environmental change, we need to understand what economic constraints and incentives exist at the local level, and how these relate to decisions to use, destroy or conserve resources. We can only do this by exploring the economy of rural households, and understanding the place of plant resources within this economy.

Thirdly, finding out these values is important for policy reasons. Often environmental values are ignored in accounting for household welfare, and so policy-makers assume they have no value. But many studies have found this to be untrue (Bojo, 1993; Cavendish, 1999a; Jodha, 1986; Godoy et al, 1993; Arnold and Ruiz Pérez, 1998; Wollenberg and Nawir, 1998). These resources matter and they matter to certain groups. Understanding these values can have a major impact on policy debates – for example, on the value of agricultural intensification; on proposals to privatize the commons; on the design of resource management systems; on the provision of emergency relief; on improving food security; and so on. But our knowledge of resource values is still sketchy.

Both economic and non-economic studies have dealt inadequately with these issues in the past. Economic studies have ignored the important contribution that natural resources make to rural households, and thereby have undervalued these resources, with sometimes disastrous consequences. Non-economic studies have tended to focus on species and resources, and have worked on valuation at that level, rather than treating resource values as part of a broader household economy. While such a focus may be appropriate for specific purposes and within the constraints of research budgets, more studies on the entirety of household income should be encouraged. To address these inadequacies, data sets are required which systematically integrate environmental resource use into the wider set of household economic activities. In designing such data collection, it is useful to follow certain general principles to make subsequent analysis and results as broad and robust as possible. Unfortunately, these principles have not been followed in many other studies (Godoy and Bawa, 1993).

The first principle is that the data should generally be household based, for reasons to be explained. To date, this household focus has been uncommon or the definition of the household unclear. Secondly, the coverage of environmental resource uses should be as comprehensive as possible. Rather than focus on individual uses or individual species, it is preferable to collect data on as many different resource uses as possible. Thirdly, data collection on resource-use values should generally be as explicitly quantitative as possible in order to generate the type of data that can be used for standard statistical analysis and comparison. Fourthly, data collection on the more standard economic activities of the household should also be as comprehensive as possible, so that a full picture can emerge of the structure of household economic activities and economic choices. Finally, given the complexities of the topic, it is useful to describe in detail the methods used. A recent review of income and livelihood studies shows that the plant income data collection methods are seldom described in sufficient detail to allow others to replicate the results or undertake comparative studies (Wollenberg and Nawir, 1998).[1]

Nevertheless, while collecting data on the full spectrum of household economic activities is desirable, it is sometimes impractical or – in the case of studies with narrower objectives – may be unnecessary. In this chapter we describe a comprehensive approach to data collection; users can, of course, select those components of the method they require for their specific purposes. Wollenberg and Nawir (1998) provide a good discussion of the trade-offs between different possible methods, including those not based on questionnaires.

What this chapter covers

This chapter aims to clarify how plant income data can be collected, cleaned, valued, aggregated and treated in the standard household economic accounting system. It is structured as follows. The next section discusses general data-collection procedures, the use of questionnaires and recall to get answers, cross-checking, the justification for using the household as the basic unit of analysis, and collecting interannual (longitudinal) data. The section entitled 'A taxonomy of environmental resource use' charts the list of environmental resource uses which exist and notes the wide range of environmental resources that households in single study areas use; the often large range of species for each use; and the breadth of economic functions that environmental resources offer to households. The section on 'Constructing environmentally augmented household income data' discusses the various procedures that need to be adopted to produce environmentally augmented household income accounts, and the classification of resources into income and expenditure. The concept of net income is introduced and the section looks at some of the problems faced in disaggregating environmental income sources from other activities.

The biggest problem, generally, when dealing with environmental resources is valuation, (conversion to a money metric). The section on 'Valuing environmental goods' describes in some detail alternative approaches, given that these resources have often been excluded from household surveys. For all environmental goods, we have the option of determining a gross value or net values of resource use, depending upon treatment of labour inputs. In the section on 'How should

labour be handled?', the argument is made for the use of the former, but data-collection procedures for labour analyses are also described; these may be needed for particular questions. The section on 'Comparing incomes across households' returns to the issue of the household as the unit of analysis and explains the use of an equivalence scale to counter problems of interhousehold variability in size and composition. The section also looks at some basic ways of exploring interhousehold variability. The final section concludes.

The case study

Much of the material for the chapter is based on two surveys on the economics of environmental resource use carried out by the author in Shindi Ward in Chivi District, Zimbabwe. These surveys comprised full household data collection in two different years (August 1993–September 1994 and August 1996–September 1997) and involved a random sample of 197 panel households in 29 villages. Shindi Ward was chosen for this work as being typical of Zimbabwe's communal areas, and indeed of many forest-dependent communities, in that it is poor; has marginal soils for agriculture; has low but variable rainfall (mean 546mm per annum, standard deviation (SD) 204mm per annum) and is prone to drought; lacks basic infrastructure such as tarred roads, piped water or electricity; has an agropastoral or hoe-based agricultural system; and relies to a degree on remittances from non-Shindi sources. It is not a resource-abundant frontier zone, having been settled for generations with substantial population growth since the

1950s from both natural increase and forced resettlement. As a result, the environmental resource base is largely reduced to communally held refuge woodlands and grasslands on mountains, hills, riverine areas and plains. Shindi has miombo and mopane woodland (Frost, 1996).

Data collection, recall, cross-checking and the unit of analysis

The data collection approach

Quantitative data are typically collected using household-based questionnaires, and this is the approach that is discussed here.[2] Some of the key issues to be considered in implementing questionnaires are given in Table 2.1. For such surveys, large numbers of households can be interviewed. Depending upon the objectives, and in some circumstances, other techniques can be of value (see Wollenberg and Nawir, 1998, for a description of such methods). For instance, data can be collected through record-keeping by local informants where more in-depth data are required and a smaller sample size is acceptable. In this method the informants record their own behaviour (Figure 2.1). Participatory rural appraisal (PRA) tools can also be used (Hot Springs Working Group, 1995; Bishop and Scoones, 1994; see Chapter 6), and indeed local PRA work is often an essential precondition for designing a successful and locally appropriate quantitative questionnaire (Richards et al, 1999a). During the PRAs, a basic understanding of the study households can be

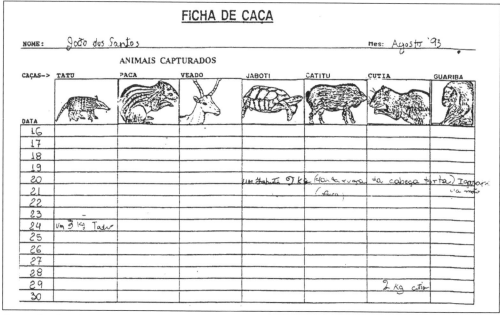

Source: Shanley, 1999

Figure 2.1 *A sample portion of a page from the notebook used by households to record daily household consumption of game*

gained so that questions posed in the questionnaire are relevant and have the appropriate responses pre-coded. Other studies have used direct observation of incomes, though in these cases sample sizes are necessarily limited (eg Melnyk and Bell, 1996).

Questionnaires are best administered in the vernacular by teams of local enumerators, ideally trained and supervised by the researcher. Many African rural areas lack an official census, so the researcher must generate his or her own household roster for the study area from which a sample of households can be randomly selected. This roster can be compiled by asking local village leaders or other authorities to name the household heads under their jurisdiction; however, these lists should be checked with other key informants and updated at the end of the fieldwork period.

A useful general template for questionnaires eliciting economic data is the income, consumption and expenditure (ICE) example (see Deaton, 1997; Grosh and Muñoz, 1995). However, income, expenditure, consumption and agricultural categories should be matched to the actual set of activities undertaken by households in the study area. This matching may be accomplished by generating a local listing of these activities prior to the questionnaire survey implementation, perhaps during PRA (see Chapter 6). The number of questionnaire rounds depends upon the aims of the researcher. If annual data are necessary, it is usual to conduct four quarterly surveys for recurrent activities, augmented by beginning- and end-of-period surveys for household socio-demographics, durables and assets, including livestock and housing. Naturally, the standard ICE survey needs to be expanded to include special sections on the quantitative use of environmental resources. Additionally, a range of special questionnaire modules may need to be added to the ICE framework to cover specific environmental uses in more detail – for example, firewood collection and storage, housing and construction, tree planting, field and environmental improvements, fencing, agricultural risk, etc. Thus some authors, such as Cavendish (1997), have used up to eight questionnaire rounds over the period of a year.

Table 2.1 *Key issues in the design of questionnaires*

1. Definition	Exactly what do you want to know?
2. Efficiency	What is the minimum you need to know?
3. Length	Keep it as short as possible.
4. Logical sequence	Try it out many times
5. The respondent's situation	Will the respondent understand the question? (Watch out for subtle shifts of meaning in translation; do not take any meanings for granted) Will she/he know the answer? Will she/he tell it?
6. Recall period	If you are asking about past events, can the respondent remember them and how do you know this?
7. Layout	Make it clear; take the space needed Design for analysis (think about coding now)
8. How will you analyse the data?	Which package? What kind of analysis? Do you need all your data?
9. Field test the questionnaire	Can you print a revised questionnaire in the field?

Source: adapted from Adams and Megaw, 1997

Using recall

To get accurate economic values, it is important to ensure that the respondent is not expected to have excessive recall of past events. Therefore, the best recall period for each item needs to be investigated locally and the questionnaires designed accordingly. Environmental resource use should normally be based on quarterly recall or even shorter periods (for quarterly recall, for example, households would be asked once per quarter what their resource use was in the last three months). This is because many uses are of a casual or short-term and therefore forgettable nature, and are additionally often strongly seasonal; as a result, attempting annual recall, or once-off fortnightly recall, will significantly understate environmental resource use.[3] Quarterly surveys also improve data in other areas, notably cash income, own consumption, gifts and recurrent expenditures. Requesting respondents to recall the previous 24 hours is also sometimes used, in which case the numbers of visits to the household should probably be increased. Thus, for example, Luckert et al (under preparation) visited households for short periods six times in each quarter to recall the allocation of labour in the day prior to the visit.

Cross-checking and setting the context

To get high-quality data, it is advisable to build comprehensive cross-checking into the research programme. For example, within-questionnaire cross-checks should be included to show up respondent inconsistency and enumerator error, whether in questioning or recording the data. Across-questionnaire cross-checks should be included, again to reduce respondent inconsistency, and also to control for the inability to interview the same household

member at each visit. Random follow-ups can also be undertaken in each questionnaire round to check the translation of questions into the vernacular and to monitor the enumerators. Perhaps most valuably, a range of qualitative information should be collected as a supplement to the questionnaires. For example, in the Shindi case study this took the form of interviews with groups of resource users, with local authorities, whether traditional (chiefs and headmen) or modern (councillors), and with local historians and elders; life history work; collection of aerial photographs; resource walks; work in archives; and species listings. As well as providing a better general understanding of many non-economic aspects of the research area, this qualitative information can also be used to cross-check the plausibility of the quantitative data appearing in the household questionnaires.

The household as the unit of analysis and handling absentee members

So far, this chapter has described collecting data at the household level. Is the household a sensible unit of analysis in rural Africa? There are two types of argument generally made against this. The first is that, due to risk, kinship ties or the 'moral economy', interhousehold transfers of goods, labour and capital reduce the importance of the individual household as a unit of either production or consumption analysis. By contrast, the second is that significant economic exchanges occur within the household, so that aggregating data at the household level results in important gender and generational issues being ignored. While both these arguments have some force, in many societies the mechanics of household formation suggests that the individual household remains a relevant unit of analysis (Box

2.1). However, the choice of the appropriate unit of analysis should be made by the researcher, according to the local social and economic system. In some cases, land areas may be the more appropriate unit of analysis (though land-based estimates can be derived through household surveys) (see Boxes 1.4, 2.2, 4.7 and 5.1). In other cases, it may be more appropriate to stratify the sample by gender groups and/or by broad kinship groups rather than by household. In making these decisions, it is always useful to draw on any sociological work that has been done in the research area.[4]

If working with households, an issue arises as to who should be a household member. The usual method of doing this is to classify a person who has spent more than six months at home as a household member, while a person who has spent six months or less at home is classified as a non-household member. Such a procedure is flawed. Firstly, the cut-off point is arbitrary: a person who is resident for less than six months may still have an impact on household incomes and expenditures. Secondly, exemptions are often made to the rule: for example, a male household head who is working away from the homestead is included as a household member. Thirdly, various categorizations remain unresolved by this procedure – for example, women who marry into the household or household residents who die during the period under review. To improve on this, one survey question could ask, for each person associated with the household, how many months she or he had resided at the household during the previous 12 months, and what were the reasons for their absence (work, education, marriage, divorce, death, etc). This can be used to calculate the 'adult equivalent units' in the household (see the section entitled 'Equivalence scale adjustments'). This approach avoids the problem of making a decision about who qualifies as a household member.

BOX 2.1 HOUSEHOLD FORMATION IN SHINDI (CHIVI, ZIMBABWE)

Household formation in Shindi, Zimbabwe, begins when a son marries for the first time. His new wife then moves in with her husband's parents, becoming a general 'servant', cooking, cleaning and washing clothes for her extended in-laws, while sharing her mother-in-law's kitchen (Cavendish, 1997). During this stage, food and labour are also shared among all household members. However, at some point – usually after a year or so, or when a younger son marries to provide a new source of female labour – the couple are instructed by the household head that they must build their own kitchen (ie they must start to fend for themselves). This is the point when Shindi people recognize that a new household has been formed, even if the couple continue to live on the same home site as other family members, because the couple have to be responsible for feeding and providing for themselves, largely independently of others. As a result, the nature of household formation in Shindi (including the fact that interhousehold transfers are rarely a significant share of total income) supports the use of the household as a possible unit of analysis. Of course, for research focused more on gender and child issues, it may be necessary to look inside the household at the way resources are allocated between household members. Such research requires different conceptual and empirical strategies. For more on this, see Haddad et al (1994) and Strauss and Thomas (1995).

BOX 2.2 FROM THE HOUSEHOLD TO THE LAND: RETURNS TO LAND OF FOREST-BASED ACTIVITIES IN SOUTHERN AFRICAN SAVANNAS

Bruce Campbell

The approaches

Throughout this chapter, economic values have been explored for households. However, in many cases it is important to quantify the economic returns to land units rather than households (the section on 'Selecting among decision criteria' in Chapter 5 considers different ways of expressing economic criteria). For example, it may be that forests will continue to be converted to arable where the returns per hectare of land under forest are less than those for agricultural use.

Four approaches have been attempted to estimate returns to land – namely, those approaches based on:

1 household studies;
2 the ecological production characteristics of the land (see Box 1.4);
3 the study of use from a 1ha patch of forest (Shanley, 1999); and
4 contingent valuation of units of land (see Box 4.7 for a description of this last method).

In the case of household analyses, if one understands the household economy and there has been no double-counting of environmental benefits, then one can sum up forest benefits over all households and calculate a per hectare value (if the area of the extraction is known). However, although the section on 'The treatment of labour costs' has pointed out that labour costs may not need to be subtracted from estimates of incomes, it is important to note that this is not the case for land valuations. When we make the jump from household incomes to returns to the land, it becomes important to subtract costs of labour. While returns to labour represent a return to people, labour inputs represent costs that should be compared between alternative land uses.

It is necessary to know which activities are carried out in which land cover types, and to know the area of the various land cover types. This kind of analysis can be rather crude or can be very detailed, depending upon the level of knowledge about household activities and about the extent of different land cover types.

Some results

An important finding is that people seem to value savannas in southern Africa much less than they do cropland (Goebel et al, 2000; Campbell et al, 2000b). There is even evidence of this higher crop value in areas of marginal agricultural productivity; this is probably related to land scarcity, to subsidies that farmers receive in the event of crop failure and to insecure land rights over forests. On the edge of Mzola State Forest, for example, where squatting in the forests is a problem due to land shortage, farmers value agricultural land at nearly US$50 per hectare per year, according to contingent valuation. Using land production approaches, the timber resources are only valued at about US$1 per hectare per year: about the same value accorded to grazing land (Gwaai Working Group, 1997; Adamowicz et al, 1997). In other poor agricultural areas in Zimbabwe, a selected range of subsistence forest resources is valued at US$5–$17 per hectare per year using household approaches (Hot Springs Working Group, 1995;

Campbell et al, 1997b; Campbell et al, 1991). Fuelwood for household cooking, wild fruits and poles are the most important contributors to this value. In a participatory appraisal scoring exercise about forest areas (see Chapter 6), extra-market values (ecological and cultural) made up 41 per cent to 61 per cent of total value. If applied to the figure in the previous sentence, this would give a total forest value (subsistence and extra-market values) in the order of US$10–$35 per hectare per year. Therefore, it may be that forests today are not profitable enough to be preserved for purely economic reasons.

Collecting multiyear (longitudinal) data

In many cases, valuation studies have been based on a single year (or less) of data collection. While this is sometimes dictated by the resources available for the research, this situation can be highly unsatisfactory. Elsewhere in this book, smallholder and forest-dependent systems are described as being dynamic and changeable. This can be due to climatic factors such as rainfall (Figure 8.4), the complex fruiting patterns of tropical trees (Box 7.6), rapidly changing economic conditions (Figure 1.4), changing land use, patterns of migration, disease outbreaks and so on. Thus data collected from a single year may not represent anything other than the specific, perhaps peculiar, conditions of that year (see last paragraph in Box 2.3).

A number of key issues need to be considered when conducting longitudinal surveys (ie surveys over term). These are briefly outlined below, and are again discussed in more detail in Deaton (1997).

Which households to survey

When resurveying households, the researcher can choose either to work with the households from the previous questionnaire round – what is called panel data – or to create a new random sample of households. Both choices have their advantages and disadvantages. The advan- tage of following the same households over time (panel data) is that it allows a clearer understanding of the relationship between life-cycle factors such as ageing, marriage, divorce, etc, and economic and environmental choices. It allows researchers to control for the multitude of household-specific factors that affect household outcomes. However, panel data are prone to the considerable measure- ment error that is germane to household data collection. Resurveying is often more expensive if the original households have to be tracked down and interviewed. And sample attrition, where households disap- pear from the sample due to migration or dissolution, can undermine the lack of bias in the original survey. The advantages and disadvantages of working with a new random sample are exactly the opposite – less can be said about household-specific factors, but the resurvey will generally be easier and cheaper to carry out. Once again, the choice between the two must be made by the researcher with reference to the aims of the research and the budget and time available.

Real or nominal prices

The existence of inflation means that the same amount of money cannot buy the same amount of goods in two different time periods, since prices have risen in the meantime. Because of this, monetary values need to be adjusted for inflation in order to reflect the real purchasing power

---

BOX 2.3 DIRECT-USE VALUES OF GOODS AND SERVICES ASSOCIATED
WITH LIVESTOCK IN THE BUSHBUCKRIDGE LOWVELD, SOUTH AFRICA

Charlie Shackleton, Sheona Shackleton, Thiambi Netshiluvhi,
Freddy Mathabela and Calvin Phiri

In South Africa many rural households either own livestock or aspire to do so. The reasons for this are complex and differ from place to place. Some communities keep cattle primarily for ploughing, others for milk and others for investment. Irrespective of the primary reasons for cattle and livestock ownership, most households derive not just one, but a number of different goods and services from their livestock. There has been relatively little attention given to all the multiple uses, especially for small stock.

We attempted to determine the direct-use values of goods and services derived from cattle and goats by rural households in the Sand River catchment of the Bushbuckridge area, Northern Province. We also estimated the benefits and costs of the presence of livestock in the area to non-owning households. Data were collected via a structured questionnaire to 101 households distributed between 13 randomly selected settlements in the catchment. The sample was stratified to include non-owners (26 households), owners with five or fewer animals (23), six to ten animals (24) and more than ten animals (28). Previous work in the area based on a random sample indicated that the mean numbers of animals per household were 3.3 + 0.5 cattle and 2.1 + 0.2 goats. Overall, only 24 per cent of households possessed cattle and 34 per cent owned goats. The ratio of mature male to female animals was 1: 2.8.

The questionnaire included sections on each possible use or service derived; amounts yielded; the proportion consumed or used by the household; the proportion sold, bought or given away; the costs of keeping livestock; and local prices for the different goods and services from other livestock owners as well as local retail outlets (Shackleton et al, 1999b). One community workshop was conducted to obtain more qualitative information and insights, as well as key informant interviews with local butchers and animal health officers. Herd census and mortality data were supported with records from the department of agriculture.

Costs of keeping livestock
The annual direct costs (excluding own labour) of maintaining livestock differed between cattle and goats (Table 2.2). The total costs per goat (R47.41) were approximately 40 per cent of that per cow (R112.38). Stock theft removes approximately 4 per cent of the herd per year. The cost to non-owners was taken as the cost of fencing their residential and arable plots (annualized over the lifespan of the fence), although many said they would fence their residential plot for privacy irrespective of the presence of livestock.

Returns to keeping livestock
Gross returns were several times higher than costs (Table 2.3), even when the savings value (via herd growth) is ignored. Direct goods and services (excluding savings value) accounted for 56.8 per cent and 41.4 per cent of the total value for cattle and goats, respectively. For cattle, the largest contributors to the total gross value were savings, milk and manure. Goats provide less goods and services, with savings, meat and cash sales being the main benefits.

Table 2.2 *Annual costs (South African rand) per owning household associated with keeping livestock during 1988–1989 (own labour not included)*

Cost item	Annual costs per household (South African rand)	
	Cattle	Goats
Direct costs		
Hiring herders	317.31	135.78
Taxes/fees	16.85	2.48
Supplementary feed	65.57	negligible
Kraal construction (annualized) and maintenance	22.32	2.44
Equipment (plough and yoke) (annualized)	6.67	–
Purchasing of stock	319.87	35.08
Indirect costs		
Losses due to mortality and theft	364.00	170.30
Total costs	1112.59	346.08

US$1 = R6.10.

Table 2.3 *Total annual value of the goods and services provided by cattle and goats to owning households during 1988–1989*

Good/service	Cattle		Goats	
	% of households	Mean annual value (R)	% of households	Mean annual value (R)
Draught	42	37.06	0	0
Draught hired out	12	26.70	0	0
Transport	15	163.72	0	0
Milk	42	883.97	2.6	0
Manure	91	763.51	0	0
Dung as a sealant	56	49.68	0	0
Dung for burning	18	73.26	0	0
Slaughtering	83	719.83	97.4	176.33
Hides	49	13.44	51.3	1.29
Cash sales	76	544.35	66.7	123.06
Herd growth (savings)	100	2487.30	100.0	425.76
Gross value				
Only direct use of goods and services (excluding savings)	3275.52		300.68	
Including savings value	5762.82		726.44	

US$1 = R6.10.

The net return per household and per animal was higher for cattle than for goats (Table 2.4). This is a reflection of the greater diversity of goods and services obtained from cattle. However, assuming the mean mass of a goat is 37.5kg and that of a bovid is 400kg, then the net return per kilogram is 19 per cent higher for goats (R1.39 per kilogram) than for cattle (R1.17 per kilogram). Non-owning households also received a net positive value of R163 per year due to the presence of livestock in the area. This was in the form of ad hoc gifts of meat and milk, free use of dung for manure and for sealing floors, and cheaper access to purchased goods such as milk, meat and draught power than if they bought them from local commercial outlets. Although the financial value of livestock products to non-owners is small, it should not be underestimated. There is little doubt that the goods and services that these households obtain from livestock represents a meaningful contribution to their livelihood strategies, especially since they tend to be the poorer households. The gifts and access to cheaper goods obtained by these poorer households not only provides a cash saving, but also reinforces familial and community relationships.

Table 2.4 *Annual net value attributed to cattle and goats during 1988–1989*

	Cattle	Goats	Non-owing households
Mean herd size (animals)	9.9	7.3	0
Gross annual value per household (R)	5762.82		726.44
231.38			
Annual costs (R)	1112.59	346.08	68.40
Net annual value per household (R)	4650.23	380.36	163.34
Net annual value per animal (R)	469.72	52.10	Not applicable

US$1 = R6.10.

The values determined through this work are effectively a snapshot of the potential value at the time. The values at both the household and catchment level are not static. At the household level, they will fluctuate according to individual management and investment decisions and the prevailing climatic conditions, particularly drought. At the catchment level, the total value will change in response to density dependent effects, such as calving success, drought and competing uses for land.

of household earnings. This can be done using the consumer price index, which measures the size of inflation over time. Such figures may be available from the national statistical offices or the finance ministries. In many cases, these figures can be problematic because they may reflect urban rather than rural consumers, or they may be dubious because of poor sampling by these national offices. Some authors have converted the local prices into US dollars, using the exchange rates applicable in the different years. While this may be the only solution when consumer price indices are unavailable or unreliable, this approach is only possible when currencies are not fixed, and even then it is unlikely that exchange rates reflect what is happening with the value of currency and consumer prices in rural areas.

A taxonomy of environmental resource use

As noted above, it is not common for standard household surveys to include questions on environmental resources. Therefore, an important preliminary question for the data analysis is to decide on a definition and taxonomy of environmental uses. Cavendish (1997) divided up household economic activities into the categories 'environmental' and 'non-environmental' (or 'wild' and 'non-wild') uses. To qualify as an environmental use, the resource must be freely provided by natural processes (ie, it can be thought of as 'Nature's bounty'). This definition usually implies that all resource uses derived from common lands are classified as environmental uses. In addition, some resources derived from private lands are environmental uses. Crops in fields are not wild, because they are intensively managed, whereas certain leaf vegetables grow spontaneously in fields without planting or weeding and are considered wild. Likewise, exotic fruit trees are usually planted at homesteads and tended carefully, whereas indigenous trees, also found at homesteads, are not: they are classified as non-wild and wild, accordingly.

Using this definition, Table 2.5 presents a taxonomy of environmental uses derived for a single study site of only 300 square kilometres (km^2) in Shindi, Zimbabwe, with an indication given of the number of different species used in each case, and the economic function or functions that each product has (Figure 2.2). Three features of this table are striking. The first is the range of environmental resources that were discovered to be available for use by Shindi households. These include a wide variety of foodstuffs; a number of non-food direct uses; a large number of uses for wood, including fuel for energy, construction materials and various implements; other non-wood tree uses such as leaf litter and livestock fodder and browse; the use of grasses, canes, reeds, etc, for thatch, mats and baskets; soil for pottery and fertilizer; and gold. This list supports the suggestion that natural habitats offer rural households a substantial set of economic values: these have, in general, not previously been considered or accounted for within the household economy.

Secondly, there is often a considerable range of species that can be used per resource type. For instance, the respondents named 47 indigenous tree species as having edible fruits. The relevance of this species diversity in economic terms depends upon the type of resource under consideration. In some cases it implies that there may be a number of different resource 'commodities' that the household can use within each overall type of environmental resource. This would be the case for wild foods, where the indigenous species are as different from each other as are better-known food types. However, there are also cases where the diversity of species for a given resource more closely resembles the existence of 'brands' rather than separate commodities. An example here would be that of firewood and construction wood, where the various named species offer the same economic service to the household, but differ in terms of their quality in providing that service.

Thirdly, environmental resources offer rural households a range of economic functions. Indeed, they are utilized in all areas of the household's economic purview: as consumption goods, as

Table 2.5 *Environmental resources in Shindi, Zimbabwe (Shindi is a Ward of about 1300 households)*

Environmental resources	Number of species per resource type*	Consumption good	Durable good	Economic characteristics of each resource type				
				Agricultural input	Other production input	Input into asset formation	Indirect or non-use value	Output (ie traded) good
1. Wild foods								
Fruits	47	✓						✓
Insects	15	✓						✓
Fish	7	✓						✓
Large wild animals	16	✓			✓			✓
Mice	u/k	✓						✓
Honey	na	✓						✓
Nuts	2	✓						
Vegetables	40	✓						✓
Mushrooms	12	✓						✓
Birds	8	✓						✓
Liquids	3	✓						
Wild fruit wine	1	✓						✓
Roasted fruits	1	✓						✓
Wild fruit porridge	5	✓						
Wild fruit butter/oil	2	✓						
Roots/bulbs/leaves	8	✓						
Wild soda	4	✓						
Wild fruit jam	1	✓						
2. Non-food indirect uses								
Medicines**	46	✓						✓
Soap and shampoo	2	✓						
Glues and lime	8	✓						
Tooth-cleaning twigs	8	✓						
Insect repellent	1	✓						
Fish poisons	7	✓			✓			
Other uses	8	✓						

3. Wood uses

Use	Value
Firewood	>50
Construction wood:	
• Hut walls	22
• Hut roof beams	28
• Hut cross beams	15
• Granary walls	22
• Granary floor	26
• Granary roof beams	18
• Granary cross beams	4
• Crop storage hut	19
• Cattle kraal	22
• Goat hut	5
• Chicken pens	17
• Crop residue store	6
• Brushwood fencing	20
• Fencing poles	16
• Live fencing	2
• Doors and door-frames	14
Scotch cart frames	3
Yokes and skeys	18
Agricultural implement handles	25
Household furniture	2
Stools	17
Plates	5
Cook sticks	23
Stirring spoons	18
Mortars	6
Pestles	15
Knobkerries	5
Carvings	3
Drums	5

4. Other tree/woodland uses

Use	Value
Leaf litter	almost all
Livestock fodder/browse	almost all
Shade at home	almost all
Windbreak at home	almost all
Rain-making rituals	5

Table 2.5 *continued*

	No.*							
Seasonal indicators	5							
Children's play	10	✓				✓		
Soil erosion protection	all					✓		
Watershed protection	all					✓✓		
5. Uses of bark								
Fishing canoes	1			✓			✓	
Ropes, fibres and string	21	✓		✓	✓		✓	
Hunting nets	2			✓			✓	
African snuff	1							
Dyes	8			✓				
Pot-firing barks	6			✓				
6. Direct uses of grass, reeds, rushes								
Thatching grass	13	✓		✓	✓		✓	
House brooms	4	✓						
Yard brooms	2	✓						
Mouse-traps	1			✓				
7. Input uses of grass, reeds, rushes								
Woven hats	u/k	✓					✓	
Woven mats	23	✓	✓				✓	
Baskets (multifarious)	u/k	✓	✓	✓			✓	
8. Uses of soil								
Termitaria	na	✓					✓	
Pottery	na	✓		✓			✓	
Hut decoration	na							
9. Other								
Gold	na							✓

na = not applicable; u/k = unknown.

* The number of species that households either used in 1993/1994 or indicated as usable. For a full species listing, see Cavendish (1997).

** The number of species used as medicines does not include the large number of species used by traditional healers (*n'anga*), only those prepared and used by the sample households themselves.

Source: Cavendish, 1997

Note: Each resource type comes with its own set of specific problems for valuation. For instance, edible insects often show dramatic variation in quantities from year to year; there may be prices for some of the kinds of poles required for hut construction but not for others; while fruit consumption is highly opportunistic and eaten in situ (eg while herding cattle). Thus, recording quantities consumed and labour inputs is not a simple task.
Photographs by Bruce Campbell

Figure 2.2 *Some of the resources that have to be valued in valuation exercises in woodlands in Zimbabwe: (a) baobab fruits* (Adansonia digitata) *and edible insects at Mbare market in Harare, (b) wood of various quantities, species and sizes for hut and storage construction (Binga District) and (c) the fruit of* Uapaca kirkiana *(Euphorbiaceae)*

durable goods, as inputs into asset formation, as agricultural and non-agricultural production inputs, as output supply or traded goods, and as indirect or non-use values. This suggests a certain degree of complexity is required when examining the interaction between economy and environment in Shindi.

Other classifications can be proposed, depending upon the purpose of the study. Thus Luckert et al (2000) divided household activities into five 'sectors': crop production, livestock production, off-farm activities (casual wages, formal employment, small-scale industries), wood-related forest activities (wood for construction and fuel) and activities based on non-wood forest products (foods, medicines, etc). In this way, the risks and returns related to investing in agriculture could be compared to forest-based activities.

Constructing environmentally augmented household income data

After collection, the data from the (environmentally augmented) questionnaires need to be aggregated to construct household total income accounts, providing the foundations for empirical analysis.[5] Certain assumptions need to be made both for environmental and non-environmental variables, which will obviously affect the final outcomes of the data analysis. For non-environmental variables, there are standard aggregation methods for household budget surveys in Africa (see Grootaert, 1982). It is sensible to use these as they will enhance the comparability and acceptability of the data results.

However, there is no equivalent literature on aggregation issues for environmental variables. This means that fairly basic problems have to be addressed when developing methods for incorporating environmental data into household income accounts. These methodological issues are discussed at some length here: it is hoped that the methods outlined will become standard practice. The first problem concerns the overall definition of income, which is discussed in the following section. The section also describes how the questionnaire data is aggregated into various income sub-components. A second issue concerns which segment of the environmental resource uses listed in Table 2.5 should be included in total income accounts. To answer this, the section on 'Classifying environmental resources in income and expenditure accounts' defines environmental income by developing a matrix of resource uses stratified by source and by use. In the section on 'The principle of net income', the concept of net income is introduced. Its calculation is described and the problems that emerge are discussed. By referring to two examples, beer-brewing and crop production, the section illustrates how the problems may be overcome.

The definition of income – the need for a broad definition

Studies using the methods described in this chapter are concerned *inter alia* with measuring the value of households' uses of a set of 'freely provided'[6] environmental goods, and setting this in the context of the overall household economy. For this purpose, an appropriate measure of the household's economic status is *total income*, that is to say *cash income* plus the *value of all non-purchased goods and services* consumed or used by the household during the accounting period. One of the key messages of this chapter is the importance of using a broad measure of household welfare. Numerous studies look at cash income alone, as if that were the sum of economic value, ignoring the role of subsistence consumption, gifts and environmental resource use.

Use of household total income, as defined here, involves three caveats. Firstly, there is no attempt to value leisure, which is why income measure is called total income rather than *full income*. The absence of a leisure valuation may lead to inaccuracies when using income as a welfare measure for interhousehold comparisons, especially when this is done for households whose intensities of labour allocation differ strongly. However, the consumption of leisure is notoriously hard to quantify let alone value (particularly for African households – see Acharya, 1982).

Secondly, income, rather than consumption, is used as the measure of household economic welfare. Consumption is often preferred in household studies because it is argued to be a better measure of welfare than income (Deaton, 1980). However, the distinction between household income and consumption in poor rural households is often not large since certain significant economic activities – the consumption of own-produced goods and of own-collected environmental goods – count as both income and consumption. As these two items alone comprise a substantial portion of economic activity in many areas, income and consumption will be of similar magnitude. Choosing between them therefore becomes less important. The third caveat concerns standardization of the income measure across households. In order to make the comparison of household incomes welfare relevant, it is important to adjust each

household's total income measure by an equivalence scale reflecting interhousehold variations in size and composition (details of an equivalence scale are given in the section on 'Equivalence scale adjustments'). Thus, unless otherwise stated, when 'household total income' is written, the text, in fact, refers to 'household total income per adjusted adult equivalent unit': this latter measure is the theoretically appropriate income construct to use for interhousehold comparisons, but it is a term sorely in need of abbreviation!

Household total income measure can be broken down in a number of ways, depending upon the objectives of the research. In many studies of rural households, market activities, non-market or subsistence activities and gifts are traditionally separated out and their magnitudes compared. In this study, a further category of environmental income is added to the list to reflect the contribution that natural resources are making to rural household welfare. Thus we can define four major sub-categories of household total income:

1   The value of cash income;
2   The (net) value of gifts;[7]
3   The value of the use of own-produced/subsistence goods;[8]
4   The value of environmental resource uses.

To establish accurately the value of environmental uses, income categories that would be classified as cash income or own production in non-environmental studies should be reallocated where appropriate. In the Shindi case study, this implied that cash-generating activities that involved either the selling of environmental goods (such as sales of wild foods, thatching grass, wood) or the simple processing of environmental inputs (such as pottery, weaving, carpentry) were

Photograph by Brian Belcher

Figure 2.3 *Traditional woodcarving from East Kalimantan, Indonesia*

treated as environmental cash income, and counted as part of total environmental income. Likewise, labour income earned in activities totally dependent upon the existence of environmental goods (such as thatching, roof mending, and carving) (Figure 2.3) was treated as a component of environmental cash income, since these income sources would disappear if the environmental goods also disappeared. Furthermore, where environmental resources were used as inputs into non-environmental production, the value of these inputs was deducted from the value of gross production and rebooked under environmental income.

A prime example of this was the use of environmentally derived fertilizers (eg forest litter) in crop production. This was deducted from the value of crops produced and added back in as environmental income. Finally, an important free environmental good to Shindi households was the provision of browse and graze for cattle and donkeys. If these were not available from woodlands and grazing areas, the household would have to purchase or grow substitute foodstuffs instead. Thus, the value of livestock browse and graze was included in the measure of environmental income and subtracted from the value of livestock produced. Such a systematic classification of income sources between the four income categories listed above is required in all studies in order to ensure internal consistency of the results.

Classifying environmental resources in income and expenditure accounts

As discussed earlier, rural households commonly use a wide range of environmental goods. When attempting to include environmental resources in classifications of livelihoods, which of these should be counted as income? This question can be answered by looking at environmental resource uses simultaneously by source and by end use. Environmental goods and services can be brought into the household in one of four ways: though collection (or harvest) from the wild; through purchase; through barter; or as gifts to the household. Likewise, they can be 'disposed' of by being consumed or used as a production input; being sold; being bartered; or by being given away. This defines a matrix of possible 'source-use' combinations, and each element of this matrix can be classified as either income or expenditure, or both. Table 2.6 illustrates this classification: note that the principles it embodies

Table 2.6 *An accounting framework for environmental goods and services*

| | | Source of environmental good | | | |
		Own-collected	Purchased	Bartered in	Gift in
Use of	Consumed or used				
environmental		Income	Expenditure	Expenditure	Income
good	Sold	Income	Both**	Both*	Income
	Bartered away	Income	Both**	Both*	Income
	Given away	Expenditure	Expenditure	Expenditure	

* The value of the 'barter in' appears as expenditure and any profit appears as income.
** The value of the purchase appears as expenditure and any profit appears as income.

are consistent with the treatment of inputs and outputs for other non-environmental variables. Using this definition, three important environmental income 'source-use' types emerged in the Shindi study: the sale of own-collected environmental goods, the consumption of own-collected environmental goods, and use of own-collected environmental goods as inputs into household production. However, studies in other regions are likely to find that different 'source-use' types are of greater significance.

The principle of net income

What it means and the problems emerging

An important general principle when constructing household income accounts is the use of *net income* (or *net profit*) when measuring the real contribution of any particular activity to overall economic welfare. These net concepts should be used for measuring the welfare accruing to households from any economic activity, be it subsistence farming, cash-crop production, trading, small-scale enterprises, crafts, wild fruit sales, firewood production and so on. Their use is important in order to avoid both double counting – for example, where an own-produced input is credited to the household twice – and overestimation – for example, ignoring when the household has paid someone else for an input. Simply put, net income (or profit) means that the value of an economic activity is measured as the total value of its output(s) minus the total value of all its input(s).[9] For example, if someone earns US$1000 in informal sector-trading, and their only input cost was the purchase of the goods they are selling, totalling US$800, then the net income from informal trading is US$200. By contrast, if they additionally incurred transport costs of US$300, then net income implies they have made a loss of US$100, so informal trading would incur a negative income in the household accounts.

Therefore, calculating net income involves summing up the cost of all the inputs used in an income-generating activity and deducting these from the value of income gained in that activity. There is one important way, though, in which economic work on rural households usually deviates from this definition. When the household has used its own (unpaid) labour in household production – for example, weeding its fields, harvesting and processing its crops, looking after its cattle, collecting environmental goods and so on – this value is not deducted from the value of output; instead, it is included in net income. So net income in this chapter means the net value to a household of an economic activity, inclusive of the own use of labour income. This issue of the treatment of labour inputs is discussed extensively in the section on 'How should labour be handled?'.

For poor rural households working in areas where markets are thin, it is more challenging to calculate net income than it is for a modern economy with well-functioning markets. This is because these households are often simultaneously producers and consumers, using crop outputs and other own-produced items as inputs into other forms of production. Also, thanks to lower levels of commercialization, input quantities and/or prices can be unclear. Take, for example, beer-brewing, where the household uses home-grown crops and own-collected firewood to make beer, much of which it may consume itself. Calculating net income implies ascribing input quantities and values for firewood and home-grown crops and for outputs not sold in the market.

On the other hand, the creation of net income data for rural households is made easier as few economic activities necessitate production inputs other than the household's labour. As mentioned earlier, it is common practice not to separate out households' own labour contributions from net income (see 'How should labour be handled?'). Nonetheless, there are some economic activities that do involve the use of non-labour inputs as intermediate inputs. Where these intermediate inputs are purchased, deduction of input costs, once measured, is straightforward.

However, when these intermediate inputs are non-marketed goods, their value must be deducted from the value of that activity's gross income and rebooked accordingly, either under the own production accounts, as gifts or as environmental income. Two good examples of this concern beer-brewing, as already mentioned, and crop production. These two examples and their problems are discussed in the next sub-sections, and a further example, livestock production, is given in 'When own-reported values cannot be used: assumptions, omissions and other techniques'.

Beer-brewing

Throughout much of Africa, local beer is brewed, usually with either sorghum or millet, but also with various wild fruits or plant saps (Gumbo et al, 1990; Bishop and Scoones, 1994). In brewing sorghum or millet beer, households use crop inputs that generally come from their own stores and firewood that is generally own-collected.[10] Thus, in order to reflect accurately the net income derived from different activities, the value of firewood inputs has to be deducted from the value of the beer-brewing output and added to environmental income, and the value of crop inputs has to be deducted and added onto the value of own-production. A consequence of doing this is that the actual returns to brewing can be reduced substantially. For instance, in Shindi, whereas gross cash income from beer-brewing in the sample households totalled Z$52,979 (Zimbabwean dollars), the total value of crop inputs and firewood totalled Z$42,362 and Z$6170 respectively, leaving net beer-brewing income of only Z$4447 (Cavendish, 1997).[11,12]

Furthermore, at the household level the deduction of input values left roughly one third of beer-brewing observations with negative net income. One possible reason for these negative values is that beer-brewing is an activity involving risk. For example, there is no guarantee that all the beer will be sold, and it is possible for the beer to spoil while it is being brewed so that negative net income is a feasible outcome. What this shows is that calculating net income can give very different results from the simple use of gross returns to measure the economic contribution of different household activities.

Calculating net crop income

The conversion of crop income from gross to net values is a more significant task, in part because crop incomes comprise a significant share of total rural household incomes, but also because cropping activities pervade the household accounts. Crop outputs can be sold, consumed by the household itself, used by the household as an input into other production, or given to others, while crop inputs can be purchased, derived from environmental resources, given to the household, or even derived from past crop production (eg seed kept from previous seasons). As a result, activities relating to agriculture appear in all the total income sub-groups (cash income, net gifts, own-production and environmental income). Therefore, moving from gross to net income for agriculture will affect all of these areas.

The basic task is to convert gross crop income to net by deducting the value of crop inputs from the value of crop production. This may seem a simple procedure, but in practice problems arise. The first concerns the timing of the data collection. Most data is collected for a period of one year, spanning a full agricultural year. For example, in an area with one growing season, data could be collected from the time of ploughing in one year, through planting and weeding and then to harvest

some 10 to 12 months later. So the input values captured in such a data set would apply to the year under study. However, for much of the year, households would be consuming, from storage, crops that were produced the previous year. Thus, there is a timing mismatch between the value of outputs and the value of inputs. Secondly, the multifield, multicrop nature of most crop production makes it difficult to assign input values to individual crops. To take one example, households usually plough all their fields in one go, so that the value of ploughing services is not easily allocated to individual plots. Likewise, fertilizer is applied simultaneously across fields and crops, rendering fertilizer values largely indivisible. (Of course, this latter issue is not a problem if all that is required is total net agricultural income per household. However, it may be that different types of households or different genders produce different crops, so that disaggregation of agricultural net income by crop type is a necessary part of an analysis of livelihood strategies.)

These problems can be dealt with as follows. To measure gross crop income, one of two methods may be used. The first is simply to measure end-period crop production and use this as the figure for gross crop income over the year in question. This has the virtue of minimizing questionnaire space and being easy to calculate, but can give a misleading picture of crop income as harvests vary so dramatically from year to year. The second, more intensive but more accurate, method is to build up gross crop income from the various uses households make of their crops during the accounting year. These various uses comprise sales of crops; own-consumption of food crops; use of own-produced crops as inputs (eg into beer-brewing); and gifts of own-produced crops to other households for either food or seed (see Table 2.7 for a complete list

Table 2.7 *Gross and net crop income in Shindi, Zimbabwe, 1993–1994*

Crop activity	Number of households involved	Total value of activity (Z$)* Gross	Net
1. Crop cash income			
Sales of high-value crops (cotton, rice, etc)	9	23,503	19,900
Sales of low-value crops (maize, sorghum, millet, nuts)	91	19,237	15,370
Sales of own-produced vegetables	113	8743	7138
2. Consumption of own-produced crops			
Own-consumption of maize	210	160,332	129,011
Own-consumption of other grains	104	27,885	23,749
Own-consumption of other staples	86	8304	6503
Own-consumption of relishes	185	20,385	16,527
3. Input use of own-produced crops			
Inputs into own-produced goods – beer sold	151	40,082	32,394
Inputs into own-produced goods – beer own-consumed	12	710	524
Inputs into own-produced goods – beer for labour hire	4	290	217
4. Gifts of own-produced crops and seeds			
Gifts of own-produced crops for food and seed	39	5775	4671
Total crop income	213	315,244	256,003

* US$1 = Z$8.09 in January 1994.
Source: Cavendish, 1997

for Shindi). The values of these four uses are summed for each household to produce the household's gross crop income: this measure captures the flow of services that the household derives from its crop production during the period in question. In a similar fashion, for each household the value of all crop inputs used in the study year can be summed to produce a measure of total crop input values. These inputs comprise the value of all fertilizers; the value of ploughing services;[13] the value of all seeds; and the value of all cash payments for planting, weeding and harvesting.

To allocate the value of crop inputs to the various sub-components of gross crop income, such as those contained in Table 2.7, a weighting scheme can be used where it is assumed that the value share of a household's crop inputs is proportional to the value share of its crop outputs. To

illustrate this with a simple example, if the gross value of maize production is 50 per cent of the total gross crop output values, and if the total input costs into agriculture are Z$62,642, as in the study, then 50 per cent of these total input costs (ie Z$31,321) can be allocated to maize production, and this figure can be deducted from the gross maize value to get the net maize production value. Repeating this procedure for all crops will result in all crop input values being deducted from gross crop output values, in a manner proportional to importance of different crops in the household's agricultural output (Table 2.7). Thus the difference between gross and net crop income will automatically be the sum of all crop input values, as indeed it should.

Where any of the crop inputs are bought (eg purchased seeds, fertilizers or hired labour), their values should be discarded (they should be deducted from the gross value of crop production, and not added back anywhere else into the household income accounts). However, where crop inputs are not purchased, their values have to be rebooked in the household accounts, since they still comprise income to the household. These are as follows:

- Inputs derived 'free' from natural environments (such as collected forest litter and termitaria soil for fertilizer) should be rebooked as environmental income because they are part of the value of nature's bequest to households. Omission of these input values would understate the value of environmental resources to rural households.[14]
- Non-environmental inputs that were gifts or transfers to the household

from other households or from the government (eg seeds and fertilizer) should be included in the calculation of the net value of gifts.

- Inputs derived from the household's past crop production (such as crop seed and the use of own-produced crops to pay hired labour) should be rebooked in the own-production accounts. (In the Shindi study this was called 'input use of own-produced goods'.)

In summary, use of the net income method for crop production means that in the household income accounts, all uses of crop outputs should be valued net of non-labour inputs, while non-purchased inputs into crop production (which qualify as household income) should be rebooked accordingly.

Valuing environmental goods

This section turns to one of the biggest problems when dealing with environmental resources – namely, valuation. By valuing environmental goods, we are attempting to measure environmental resource use with a yardstick (monetary value) that allows comparison with other household economic activities. Not only is the valuation of environmental resources controversial, it also often proves difficult. This is not the place to discuss the pros and cons of valuation; rather, the purpose here is to propose a series of approaches that are usable should researchers wish to do this. Thus, the following section describes an approach to valuation which relies on households' own-reported values for environmental goods, derived from specific resource use questions on quantities used and estimated total values. The

section also indicates what can be done to check such results and gives approaches for various smaller items that prove difficult to value (eg some wild fruits, medicines). While this method provides coherent values for many environmental uses, there are some which cannot be accounted for this way and for which special valuation procedures are required, or which have to be omitted (see the section on 'When own-reported values cannot be used: assumptions, omissions and other techniques' and Chapter 4).

Using own-reported values as the basis for valuation

The method described here uses own-reported values for environmental resources. This implies one immediate omission: the method cannot be used to estimate the non-use values of environmental resources, such as the watershed protection function of forests, the windbreak function of trees, the aesthetic benefit of forests and so on. Such services require specialized valuation techniques, such as hedonic pricing, contingent valuation, travel costs methods, etc (see Chapter 4).

An important advantage of basing valuation on observations of actual behaviour is that the resulting value estimates are derived from true household choices, facing prices that are 'real', in that they reflect local demand and supply conditions. Rooting resource valuations in actual household behaviour surmounts some of the weaknesses of other resource valuation studies. For example, although the widely cited Peters et al (1989) did use

actual prices, the paper based its resource use assessment on potential rather than actual forest production and consumption (see Box 1.4).

Even if actual prices and quantities are used, an important consideration is how robust such values are to alternative use levels. In many situations, if the entire forest supply came on the market, prices would decrease, and current prices are therefore quite inappropriate for the task. The problem here lies in the use of data derived from one market equilibrium that may not apply to another simulated scenario. One way around this problem is to estimate how prices might respond to changing use levels. As an example, variable price and cost information was used for estimating firewood values in Campbell et al (1997b) (Figure 2.4).

Photograph by John Turnbull

Figure 2.4 *Firewood being transported for home consumption and sale by women and children, near Hanoi, Vietnam. Firewood is often relatively easy to value because in most countries it is usually highly valued and traded, at least locally*

Eliciting own-reported values

Own-reported values for environmental resource uses can be elicited in the same way as values are elicited for other more 'normal' economic goods. In other words, at each point in the questionnaire where households claim to use environmental resources, they can be asked to provide an estimate of the total quantity and total value of that resource use, and these responses can then be used to value environmental goods. From these individual household responses, area-wide estimates of prices and unit values can be made and applied to households who, for some reason, were unable to value their resource uses – what is known as value imputation. The potential difficulty with this as a valuation method is that many environmental goods and services are not traded in formal markets – this is one reason why they have been excluded from household budget surveys in the past. Given this absence of formal markets, it could be argued that relying on actual market prices for valuation, as own-reported values do, is flawed. As a consequence, two problems might emerge:

1   Respondents might be unable to value their resource uses, leading to a large number of missing values in the questionnaires.
2   Respondents might make up values to fill gaps in their answers, resulting in meaningless and misleading figures.

Nonetheless, contrary to expectations it is often possible to derive meaningful values for the overwhelming majority of environmental resource uses (thus, for instance, Cavendish, 1997, derived values for some 50 individual environmental goods) (see Figure 2.2). This is because, although environmental goods are not often traded in formal markets, they are generally traded or bartered locally, so that it is often easy to discover a coherent local schedule of prices and quantities.[15] Three classes of goods can be distinguished in the Shindi study. Firstly, there are some environmental goods – such as wild meats, wild fish, wild fruit-based wine, mice, household furniture, household utensils (ie cook sticks, mortars, pestles, etc), agricultural implements, thatching grass, woven mats and baskets, pottery and gold – that are widely traded locally, so much so that they effectively have a local market. For these resources, valuation is a straightforward procedure, in that households have no difficulty in assigning resource values and we simply accept the households' own estimates of the value of environmental goods consumed or used.

Secondly, there are also a sizeable number of environmental goods whose prices seem to be universally known despite absent or very thin trading. Cases of this in the Shindi study include wild foods such as insects, honey, birds and mushrooms; leaf litter, cattle manure and termitaria soil inputs into agriculture; firewood; and wood inputs into construction (Cavendish, 1999a). When asked the value of the quantity of these goods that a household was reported to have used, very few households were unable to respond. Households' stated values could therefore also be used to value these goods.

Thirdly, there are wild goods which households find it difficult to value directly, but which have clear and close substitutes. The most obvious case here is wild vegetables, of which in the Shindi study some 40 different species were reported as being consumed at one time or another (Cavendish, 1997). Very few of these species were ever traded. However, in general these species are green leaf vegetables, used as cooked relishes for *sadza* (a stiff porridge made from maize meal). Thus, they are close substitutes with each other and with domesticated species such as rape and cabbage, all of which are sold at a set price per bundle. This, then, can serve as a reasonable imputed price for these environmental goods.[16]

There are some resources, such as wild fruits and wild medicines, which are not traded and/or which do not have substitutes, so that it is difficult to use households' valuations. In some cases a modified method of using own-reported values may be used (see below); in other cases entirely different methods are needed (see 'When own-reported values cannot be used: assumptions, omissions and other techniques' and Chapter 4). However, for the overwhelming majority of environmental resource uses that one wishes to measure, households often have estimates of resource use values, and it is these that we can use.

Checking own-reported values: dispersion in the unit values of environmental goods

As stated, the valuation method adopted relies heavily on households' own estimations of resource values. However, rather than accept these at face value, it is important to ask whether these estimates are at all plausible. As suggested earlier, it may be that households feel under pressure from enumerators to value their environmental resource uses, and in consequence give answers, but that these values are largely arbitrary. Alternatively, households may regard the idea of valuing 'free' environmental goods as unusual or absurd, and therefore may not take the same care over answering accurately as they might for other goods, such as crops or livestock. If this is so, then the method proposed here – of taking household valuations as a reasonable measure of the true value of environmental resource use,

and using these for imputing missing values – will be distinctly suspect.

We can check this by examining the aggregate dispersion in the own-reported unit values of environmental goods, where unit value equals total value divided by the number of units used. If households are indeed responding as suggested above, it is reasonable to expect a high degree of randomness in individual household responses, and this should show up in aggregate as a high degree of dispersion in the implicit unit values for environmental goods. By contrast, if the valuation method is a reasonable one, one should find aggregate unit value data that look usable as prices, in that each environmental good has a clear implicit price, and price differences within goods and across goods relate systematically to differences in quantity and quality. To illustrate this, a portion of the data from Shindi is displayed (Table 2.8), showing some basic distributional statistics for unit values. Note that for some of the environmental goods there are several different units of measurement in common use for assessing the quantity of the resource used. In these cases, unit value statistics are given for more than one unit of measure.

On the basis of these distributions, it would appear that the household's own-reported valuations may, indeed, be used as a good measure of the value of environmental resources. Various features of Table 2.8 support this assertion. Firstly, barring a small number of exceptions, the standard deviation of each unit value distribution is less than the mean, and in many cases is less than half the mean. In other words, the own-reported unit values are quite closely clustered around the mean, not the pattern one would expect if individual households were answering questions in a random manner. Secondly, for almost all of the unit values listed, the mean and median unit values are very

Table 2.8 Some implicit unit values of environmental goods in Shindi, Zimbabwe, 1993–1994*

Environmental Resource Use	Unit of measure	Number of observations**	Mean	Standard deviation	Implicit Unit Value (Z$)***			
					Median	Mode	Minimum	Maximum
1. Wild vegetables								
Phaseolus vulgaris (fresh)	20 litre bucket	93	9.92	7.73	10.00	10.00	1.00	40.00
	crop basket	71	0.85	0.27	1.00	1.00	0.50	2.00
Phaseolus vulgaris (dried)	20 litre bucket	164	22.61	7.24	24.00	20.00	6.00	40.00
Cucurbita pepo (fresh)	20 litre bucket	57	8.45	3.30	10.00	10.00	1.00	20.00
	bundle	285	0.94	0.18	1.00	1.00	0.20	2.00
Cucurbita pepo (dried)	20 litre bucket	29	22.00	7.32	20.00	20.00	10.00	45.00
2. Wild fruits								
Sclerocarya birrea fruit	20 litre bucket	13	1.93	0.85	2.00	1.00	1.00	4.00
Sclerocarya birrea nut	50kg bag	26	14.25	9.87	10.00	10.00	3.00	40.00
	20 litre bucket	99	5.66	3.45	5.00	5.00	1.00	20.00
Berchemia discolor	1kg sugar bag	33	0.68	0.34	1.00	1.00	0.10	1.00
Artabotrys brachypetalus	1kg sugar bag	18	1.05	0.48	1.00	1.00	0.05	2.00
Ximenia caffra	1kg sugar bag	23	0.86	0.59	1.00	1.00	0.10	2.00
Vanguaria infausta	1kg sugar bag	13	0.80	0.47	1.00	1.00	0.25	2.00
Flacourtia indica	1kg sugar bag	11	0.52	0.33	0.50	0.25	0.25	1.00
Syzygium guineense	1kg sugar bag	4	0.85	0.30	1.00	1.00	0.40	1.00
3. Other wild foods								
Impala	meat bundle	33	2.10	0.39	2.00	2.00	2.00	4.00
Kudu	meat bundle	6	2.17	0.41	2.00	2.00	2.00	3.00
Zebra	meat bundle	12	2.00	0.00	2.00	2.00	2.00	2.00
4. Firewood								
Firewood	scotch cart	236	10.42	4.28	10.00	10.00	2.00	30.00
	bundle	1608	2.02	0.80	2.00	2.00	1.00	7.00
5. Small and large carpentry items								
Cook stick	one	66	1.20	0.71	1.00	1.00	0.20	5.00
Cook stick	one	58	1.19	0.49	1.00	1.00	0.25	2.00
Cook stick	one	42	1.05	0.44	1.00	1.00	0.30	2.50
Cook stick	one	25	1.39	0.73	1.00	1.00	0.50	3.00

* The underlying data were derived either from the quarterly questionnaires on resource use, or from one-off data collected on the value of durables and livestock numbers and values.

** This refers not to the total number of observations of a resource use, but rather the number of observations with meaningful value and quantity data for a given resource use and unit of measurement.

***All data are from 1993–1994; US$1 = Z$8.09 in January 1994.

Source: Cavendish, 1997

close in value, and it is common for the modal and median unit values to be the same figure. Thus, not only is there little skewness in the household's unit values, the most popular own-reported unit value is also generally one of the measure of central tendency.

Thirdly, where an environmental good is measured with more than one unit, the ratio of reported unit values generally matches the ratio of quantities, as prices should. For example, firewood is commonly transported by head in bundles or by scotch cart. Local respondents reported that, on average, five bundles of wood could fill a scotch cart: this ratio of quantities is matched by the ratio of both mean and median unit values for the two firewood units. Fourthly, where similar environmental goods are measured by the same unit, the reported relative unit values make sense. Thus, across a variety of wild fruits species, the unit value of a plate of fruit is equal to the unit value of a 1kg sugar bag, and both generally equal Z$1. Similarly, although there are four different types of cook sticks, each with a different function when preparing the staple *sadza*, they are of similar size and shape, so that one would expect them to have very similar unit values. This is indeed the case. Other examples include the unit values of the same unit of different wild vegetables and the same unit of wild animal meat. Finally, where foods can be either used fresh or prepared in some way, we would expect the unit price of prepared foods to be higher than that of fresh food, given the extra labour involved. This is what is found. For the wild vegetables *Phaseolus vulgaris* and *Cucurbita pepo*, in both cases the unit value of a 20-litre bucket of dried leaves is roughly twice that of a bucket of fresh leaves. Similarly, the unit value of a 20-litre bucket filled with nuts from *Sclerocarya birrea* is more than twice the unit value of a 20-litre bucket of the original fruit from which it is priced.

This suggests that despite occasionally thin markets, environmental resource uses do have recognized unit values, and these unit values are interpretable in an economic sense as prices. This supports the valuation method proposed here, which accepts households' own estimates of environmental resource-use values as accurate measures of the welfare these resources generate.

However, there is still a fair degree of dispersion in the unit values for environmental goods presented in Table 2.8. This can be seen in the ratio of maximum to minimum reported unit values for each good, which in many cases is quite high. This high ratio, or equivalently the presence of extreme unit value outliers, might suggest that certain households were very mistaken in their evaluation of resource values. To deal with such unit value outliers, the researcher can adopt a procedure whereby an imputed price (such as the median of the distribution) is applied to these households in order to revalue their resource uses. In other words, for most households we accept their own valuation of resource values, but for others we may adjust their estimates if we believe them to be unreliable.

Nevertheless, there are good reasons why one may choose not to adjust unit value outliers in such a fashion. Although respondent error is germane to household questionnaires, there are genuine reasons why the reported imputed unit values in Table 2.8 should differ, even when the underlying prices of goods are identical. One reason is that the units of measurement are, in practice, not identical. This is natural for goods that are measured in single units of a species, such as mice, birds and fish. Quite obviously these will vary in size and so will the total value, resulting in a variation in imputed unit values to match the variation in size. However, there are other units of measure which also are not standardized. Bundles of firewood and bundles of thatching grass can both vary in size greatly, as can many other measures. These size variations – unrecorded in questionnaires – will result in genuine variations in implicit unit values.

Another reason for dispersion in the implicit unit values is that of (uncaptured) quality differences. Quality differences are as marked for environmental goods as they are for other standard goods. In the Shindi study, for most of the goods for which producer group interviews were held, quality differences were clear and understood – for example, in items such as firewood, construction wood, other construction materials, pottery, implements, furniture, thatching grass, mats, baskets, wild foods, etc (Cavendish, 1997). Along with this went general knowledge that the top craftspeople of the area – the best carpenters, potters, builders, thatchers, weavers, hunters and so on – could command higher prices for their goods and services. Most implements and utensils differ in the quality of their constituent materials (for example, better or worse woods for carpentry goods, clays for pots, and rushes/canes for baskets and mats) and the skill with which they are made, so that their prices will also vary accordingly. Another quality-type problem that will lead to variations in the unit values is depreciation. This is most relevant for the non-food environmental goods, especially those with durable characteristics, such as large carpentry items, mats, baskets and construction materials. Households could be asked to value their stocks of these goods at current resale prices. Their estimation of this resale price will naturally be affected by the age and condition of the good. As a result, we should expect the unit values of

environmental goods defined only in quantity terms to display some degree of dispersion in response to these variations in quality.[17]

Finally, dispersion of unit values may arise from rural questionnaires because prices are not established at a standardized place – that is, there may be high spatial variability. Since transport costs figure highly in the costs of production (eg if you buy firewood from somebody at their house, you still have to carry it home), then the actual price of goods to the household can genuinely vary. Once again, this will lead to variation in stated unit values across households.

In conclusion, then, in the Shindi study outlier values were not adjusted using median prices, for the reasons cited. However, this is a matter on which individual researchers need to draw their own judgements, looking at the degree of dispersion in own-reported values and concluding whether this seems plausible in economic terms or not.

Difficulties with some environmental income sources

There are some wild goods for which own-reported values are scarce. These are problematic cases, and in each instance an appropriate valuation method needs to be developed. Examples of problem cases from the Shindi study are highlighted below (Cavendish, 1997). The most significant problem was wild fruits, of which 47 species were reported as being consumed during 1993–1994; however, with very few exceptions, such as *Sclerocarya birrea*, *Strychnos cocculoides* and *Strychnos madagascariensis*, they were neither purchased nor sold. Furthermore, their substitutes (exotic fruits such as mangoes, oranges and bananas) are much higher-quality foods, so that the prices of these could not be used to impute values for wild fruits. Nonetheless, although many wild fruits were neither sold nor bartered, respondents were asked as for other environmental goods to estimate the quantity (using self-chosen quantity units) and total value of the fruits consumed. Remarkable consistency was found in the relative values assigned to different quantity measures of different wild fruit species (Table 2.8). Thus, for almost all of the wild fruits for which respondents estimated quantities and values, one normal-sized plate of wild fruit was given the same value as a 1kg sugar bag of wild fruit and both were given twice the value of one cup and one tenth the value of a 20-litre bucket. The consistency in respondents' relative valuations across households and across different wild fruit species allowed a chain of wild fruit values to emerge and, hence, value imputations to be made for observations which otherwise would have had missing values. Ultimately, this allowed almost all wild fruits to be assigned a use value.

There was a similar problem in valuing wild medicine use; 46 wild species were reported as having medicinal use, though, in fact, only 26 of these were reported as actually being used during 1993–1994.[18] The author suspects, though, that information on wild medicine use is a substantial underestimate. Respondents were sometimes embarrassed to report use of these medicines because they were perceived as being inferior to 'Western' medicines available at the local clinic. Likewise some local evangelical churches regarded the use of local medicines as 'ungodly' and ordered people to desist from using them so that respondents were likely to hide their true level of use. But the main problem came with valuation. The only purchased traditional medicines are those bought from a traditional healer (*n'anga*). However, the 'service' being paid for mostly consists of

the *n'anga*'s divination skills and his/her ability to intermediate with the spirit world, rather than the medicine used (see Gelfand et al, 1985). Although the household's own use of traditional medicines is mostly for illnesses for which there exist locally purchasable 'Western' medicines (for example, flu, diarrhoea, other stomach complaints and eye infections), they are not truly comparable since the efficacy of the traditional medicines is in question and quality differences are probably large. The solution to these problems resembled the case of valuing wild fruit. Households were again asked to provide a subjective valuation of the resource usage, and these valuations were compared. As for wild fruits, a certain regularity emerged in household valuations – namely, for frequently used wild medicines, a mean value of Z$2 was reported, whereas for less frequently used wild medicines, a mean value of Z$1 was reported. These mean values were then used to value those cases where the households had not volunteered any valuation information.

When own-reported values cannot be used: assumptions, omissions and other techniques

In some cases, there are environmental resources for which no own-reported values can be obtained. One may need to make assumptions, explicitly indicate omissions or turn to other methods to value the resources (see also Chapter 4 where specialized valuation techniques are described). In the following sub-sections particular examples are raised.

Making assumptions – the case of toothsticks

In the Shindi study, eight plant species were used to clean people's teeth and freshen their breath. The twigs were picked opportunistically or from trees near the homestead and were not traded. The only substitutes are purchased toothbrushes and toothpaste, but the quality difference between these and toothsticks is enormous. The only option was to make an arbitrary assumption that the value of using toothsticks was Z$0.05 per day. As with wild medicines, the aggregate value of toothstick use was very small, so this assumption makes little difference to the household income accounts. Furthermore, interviews with a range of local respondents made it clear that toothsticks were of very minor importance when it came to the total value of resource uses.

Admitting defeat – the case of water

There are some values that prove hard to measure – for example, the value of water consumption. Measuring the quantity of water used for drinking and cleaning is often not problematic. However, assigning a value to this water can prove near to impossible. Water can be one of the few economic goods for which no sensible price can be derived either from formal questionnaires, from qualitative data collection or from comparison with a backstop or substitute resource. In addition, water quality often varies hugely. Boreholes have the highest-quality water, but the water from natural sources may vary in quality with respect to its muddiness, its taste and/or its infection, meaning that even if it were possible to derive a price for borehole water, this could not be used to value other water sources. In such cases it may be best to omit the environmental resource from the household income accounts or use other methods (see Chapter 4).

Changes in the stocks of renewable resources

If the income measure used is to be a true measure of sustainable income, we need to adjust the value of the household's environmental resource uses to reflect changes in the stock of renewable resources (see Dasgupta, 1993, pp297–304, for a formalization of this point). The sign of the income adjustment depends upon the sign of the resource change: if the resource stock is being depleted, measured income should be reduced in order to reflect this diminution of the natural capital stock. However, this type of renewable resource adjustment is challenging and has rarely been attempted. Firstly, there is often no data on the changes in resource stocks at all. Note that given the multiplicity of environmental resources often used by households, data on resource change would need to encompass all of the underlying renewable resources generating environmental supplies, so that data would be required on forest extent, composition and structure; animal and fish species numbers; grassland extent and quality; household and field tree numbers; and so on. Such data are hard to collect. Secondly, even if data on resource stocks were available, these changes occur not at the household level but at the community level. Consequently, there is the task of converting data on general resource change into household values. In general, then, it is very difficult to attempt a valuation of stock changes at the household level.

So, where changes in the resource stocks studied are known to be small – as was the case in the year comprising the Shindi study – then the effort required to adjust household accounts for changes in resource stocks is probably excessive. However, when resource change is very dramatic (eg the decline in sandalwood shown in Figure 3.4), then some adjustments are necessary to derive figures for sustainable income. Resource accounting assessments are discussed again in Chapter 5.

Dealing with livestock – the valuation of browse and graze

Livestock are often critical to household livelihoods and, in addition, are usually major users of environmental resources via their browsing and grazing (Barrett, 1992; Scoones, 1992), though in some farming systems (eg Kenyan highlands) pen-feeding using own-produced or purchased inputs is more common (Figure 2.5). In most of the savanna region, browse and graze are freely provided to livestock-owning households by the environmental resources of communally held grazing areas, and their value can be expected to be substantial since they are often the majority of the fodder inputs to large livestock (eg cattle, goats and donkeys; the remaining fodder inputs are comprised of crop residues). However, valuation of these fodder inputs is not easy. There is often no market for fodder inputs, and households are usually unable to place values on them directly. Although substitute livestock feeds do exist, they are rarely used. Furthermore, because they are often prohibitively expensive to use, their value cannot be employed as a proxy for the value of the resources that livestock actually consume. In addition, it is not usually possible to calculate the value of livestock fodder and browse using land markets. Land is rarely rented for grazing purposes and, in many cases, is even not rented for cropping purposes. Thus, fodder and grazing values can often not be directly calculated at the input end, or 'point of ingestion'.

Given this difficulty, it makes sense to attempt a valuation of livestock feed at the output end – namely, by evaluating the contribution that livestock make to the household economy. Two options for this are possible. The first is to value and sum specific livestock outputs per household per annum (eg manure, draft power, milk, meat, change in stock values, etc). The problem with this option is that some of these will be measured very badly in the questionnaires, especially cattle values that are not agricultural inputs (eg pulling scotch carts for crop deliveries, fetching firewood, transporting construction materials). A number of studies have been conducted on the value of cattle to the production system, and the results of these studies vary hugely both in absolute and relative terms (see, for example, Campbell et al, 2000a; Scoones and Wilson, 1989; see also Box 2.3). The differences may be partly due to differences in the livestock systems, but are more than likely a result of differing assumptions about livestock productivity parameters and of differing methods of questioning.[19]

The second method, proposed in Cavendish (1997), is to exploit the fact that in an efficient market, the current price of a livestock unit should reflect the net present value (NPV) of its entire future income stream to the household (see 'Selecting among decision criteria' in Chapter 5 for a discussion of NPV). Assume, for example, that cattle or donkeys are a 'frictionless' production technology, merely converting food inputs into useful outputs at 100 per cent efficiency and with no value added in and of themselves. With this assumption, it becomes possible to use (own-reported) livestock prices to calculate an income stream over time that will equal the value of all fodder inputs. If the income stream from livestock is evenly distributed over time, the current year's livestock income stream (y_0) is given by:

$$y_0 = P_0 \left(r / \{ r - (1/(1 + r))^T + 1 \} \right) \quad (2.1)$$

where T = lifespan of the livestock unit measured from the current date

r = the discount rate (expressed as a proportion)

P_0 = the current price of the livestock unit, based on own-reported values

In the Shindi study, we assumed that T equals five years and that the rural household's discount rate was 10 per cent.[20] The calculated income stream y_0 is then, by assumption, equal to the total value of feed resources in the current year. The value of the livestock environmental resource inputs is then equal to the proportion of livestock diets that comes from browse and graze, rather than other sources, such as crop residues or purchased feedstocks. There may be a problem with obtaining this proportion, but it may often be gleaned from the range land and livestock literature (eg Campbell et al, 1998).

Note that some quite demanding assumptions underpin this procedure. Firstly, the income stream derived from cattle and donkeys is not equally distributed over time. As young animals, their contribution to income is likely to be negative; as they grow older and stronger, their contribution to income will increase until, beyond a certain age, they will be less powerful and hence less able to provide draught power, so that their income stream will start to decline. Secondly, the value of cattle does not diminish to zero at the end of their life because they can be slaughtered and eaten and the hides used or sold; thus, they are an asset with salvage value.[21] Thirdly, the lifespan of a unit of livestock is not fixed but highly variable, as livestock in most subsistence sectors are affected by drought, disease and theft. Thus, T is an expected variable that is quite likely to vary from one livestock unit to another.

(Given these assumptions, and the problem of picking a discount rate, this method of valuing livestock has serious weaknesses. Clearly, much more work is required in this area in order to arrive at a fully satisfactory method of valuing livestock inputs in the absence of equivalent purchased inputs or grazing rental markets. This work could involve refining the method described here, of working backwards from livestock prices, and also more systematic work on valuing the different goods and services that livestock offer rural households.)

The above procedure means that we are classifying all large livestock output values to the household as environmental income. This, therefore, requires an adjustment to the household accounts, in that:

- The sales of large livestock and their produce should be excluded from household total income, since these have already been included in the calculation of total livestock income.
- The contribution of livestock as an input to crop production should now be deducted from crop income.

The full value of livestock fodder inputs has already been estimated by calculating the current value stream associated with their asset price. As a result, if the crop input values of large livestock were not deducted, there would be double-counting in the accounts. For agropastoral households, the chief inputs are manure for fertilizer and the provision of ploughing services. So, for the income accounts to be consistent, the value of manure use and of ploughing inputs must be calculated and then deducted from the gross value of crop output as part of the derivation of net crop income (see the preceding section, 'The

Note: The valuation of fodder from such systems is usually relatively easy. However, where graze and browse are derived from free range in the dry forests and woodlands of Africa, valuation is much more difficult and may be based on the valuation of cattle rather than their feed (see the section on 'When own-reported values cannot be used: assumptions, omissions and other techniques' and Box 2.3).
Photograph by Francis Ng

Figure 2.5 *A farmer in central Java collects fodder for penned cattle from an agroforestry system*

principle of net income'). Where the household has used its own livestock for ploughing, the value of ploughing inputs must then be discarded from the accounts on the grounds that this value is already included in the value of livestock browse and graze. However, where the household has received ploughing services for free from another household, this represents a gift from natural habitats that is not accounted for anywhere else as household income (the cattle being used for ploughing are not from the household in question and therefore have not been valued). This item could be included as environmental income under a title such as 'Imputed value of ploughing gifts'.

How should labour be handled?

A key feature of rural households is that many of their economic activities – particularly the collection and processing of environmental goods – involve the use of labour, whether from the household itself or hired in from outside. To meet the net income/profit criterion, strictly speaking labour costs should be deducted from all household activities. However, as noted earlier, in practice many household studies, including the Shindi one, calculate net income inclusive of own-labour costs. To explain this, the following section discusses the difficulties of getting accurate measures of rural households' labour costs (essentially, the total time allocated to an activity multiplied by the value of that time). Researchers may therefore choose not to adjust the household accounts for labour input costs, especially since this is rarely done in other economic studies of rural households. However, should there be a need to calculate labour costs, the section on 'Surveys of labour' offers some suggestions.

The treatment of labour costs

The methods described above can be used to establish the value to sample households of the entire range of environmental goods charted in the questionnaires. However, these are still gross values in the sense that they include the own-labour costs of collection, processing and marketing. Should we go a further step and refine this measure by calculating the net value of environmental resource uses, where these involve labour inputs by the household itself?

Examining this question from a slightly different perspective, in economic studies of environmental resource use there are usually two different natural resource values that researchers may wish to measure. The first is the household income associated with the use of a 'free' natural resource. The second is natural habitat value (NHV), which is the pure value of the resource itself. NHV differs from household income in that it has costs of capital and labour subtracted, while household income does not. For most peasant uses of environmental goods, capital equipment is unnecessary for the processes of production or consumption. Thus, in general, household income equals NHV plus the labour costs of collection, processing and marketing. For completely consistent household accounting, the appropriate value to use in calculating environmental income is NHV, with the value of labour inputs booked separately under returns to household labour.

However, NHV is difficult to calculate due to the multiple problems of measuring and valuing labour inputs. To begin with, there are enormous problems in measuring the quantity of labour inputs. For example, firewood collection is sometimes opportunistic (especially when involving collection for daily cooking and heating) and sometimes planned (eg when collecting for a major use such as beer-brewing). In the latter case, time taken is a rough measure of labour input. In the former case, where firewood is 'collected' on the way back from work in the fields, 'time taken' has little conceptual meaning because no additional time has been used for this task, and yet the collection activity still has value. Furthermore, in a multicollector household, getting such data from a questionnaire is difficult. Similar problems affect almost all resource uses. Firstly, the variance of time taken per task is substantial across different users (eg old carpenters take much longer to make the same item than younger ones). Secondly, labour input is often sporadic and stretched out (eg a weaver making a basket may take months, picking up the work in spare moments and discarding it after a short time). Thirdly, a task may be multiperson and yet have the number of individuals involved change regularly (eg wood collection). Fourthly, the same task may involve differing quantities of labour input for the same person on a case-by-case basis (eg firewood collection, as well as cattle herding, weaving, insect, mouse and bird catching, etc).

There are also problems in measuring the value of labour time. There are, for example, well-known difficulties in deriving a shadow price of labour because the price varies *inter alia* according to season, age, sex, education, effort and the quantity of labour available to the household (Cavendish, 1997; Melnyk and Bell, 1996; Wollenberg and Nawir, 1998). Allocating a wage rate in a methodologically satisfactory manner requires information on every element in this potentially large matrix of labour types. Furthermore, in contracts involving payment for labour services (eg for weeding, harvesting, planting, building, thatching), closeness of relation enters as a factor in determining payments. Using these observed wage rates to proxy for the value of labour time in other circumstances is then not valid (Cavendish, 1997). Therefore, allocating a value to a unit of time for a given task is as challenging as discovering the quantity of time taken for that same task.

The difficulties of calculating the quantity and price of labour inputs in the collection and processing of environmental goods are exactly the same as those underlying the calculation of labour inputs into other productive activities, in particular agriculture. Given the low asset levels of most peasant households and their consequent reliance on labour as the main factor of production, this problem spans a whole swathe of activities. If an accurate calculation of labour costs is not feasible, then two options remain: either to make a set of arbitrary assumptions about the quantity of labour inputs in various productive activities and the associated shadow price of labour, or to calculate household accounts based on a gross (ie inclusive of labour costs) rather than a net value basis.

In general, the latter procedure is recommended. Working with household income/profit figures which include the household's own-labour input costs is not perfect in an accounting sense, but it can often save researchers an enormous amount of time and questionnaire space by freeing them from the need to investigate in detail labour allocation and labour value. Furthermore, almost all other standard economic studies of rural households also take this position. For example, the author is not aware of any major rural household study that excludes the value of the household's own labour inputs into agriculture from the value of agricultural income. In other words, working with household accounts that do not separate out the household's use of its own labour inputs in its various economic activities is both easier and also facilitates comparison of the data results with other more traditional studies.

Surveys of labour

It may be, though, that calculating environmental values net of labour inputs is a necessary part of the data analysis. This would certainly be needed if the research were aimed at calculating pure resource values (otherwise called *resource rents*) rather than household income. In this case, data must be collected on both the time spent on various tasks as well as the value of labour time (ie the wage rate). Godoy and Lubowski (1992) and Wollenberg and Nawir (1998) suggest the following methods for assessing time allocation:

- direct observation (with resultant small sample sizes);
- record-keeping by informant households; and
- interviews about the activities of the household, ideally at the end of the day.

Luckert et al (under preparation) used a technique whereby households were asked about their allocation of activities in the previous day, with households visited six times in each quarter of the year. In this way, interviews took less than ten minutes, but large numbers of households could be covered each quarter (250 in their case). Alternatively, where activities across households are fairly undifferentiated, key local informants can estimate the time taken for various activities (Richards et al, 1999a).

A value for labour can be obtained in a number of ways (Wollenberg and Nawir, 1998). Where people receive a wage for the activity, the matter is simple because this wage can be used to value labour.

However, households do not generally pay their own members a wage for their work in household production, so imputation of an equivalent wage rate is required. Among the techniques used to value labour are the following:

- the use of national minimum wage rates (Peters et al, 1989; see Box 1.4);
- the use of local wage rates for various activities (Hot Springs Working Group, 1995); and
- the calculation of wage equivalents where work is done for goods or accommodation provided (Godoy et al, 1995).

Comparing incomes across households

Part of the purpose of collecting household data is to be able to compare the empirical results of the study across households. However, it is inaccurate to do this unless the household total income figures have been adjusted to account for the considerable variability in size and demographic make-up that occur in almost all societies. Put simply, an annual total income of Z$10,000 to a household of a single adult person is not the same as Z$10,000 to a household of several adults; this is the problem of household size. Likewise, an annual total income of Z$10,000 to a household of four adults is not the same as Z$10,000 to a household of one adult and three children, due to the different needs of adults against children; this is the problem of demographic composition.

Thus, if total income is to be used as a measure of the household's welfare, then interhousehold variations in household size and demographic composition must explicitly be taken into account. A failure to make such adjustments will lead to

highly misleading results. For example, the wrong households may be found to be poor; the relationship between environmental resource use and poverty is likely to be misunderstood; female-headed households may be found to be less poor than they really are; and so on (Deaton, 1997; Deaton and Paxson, 1998; Lanjouw and Ravallion, 1995). Such interhousehold adjustments are made using an equivalence scale, which converts households of different size and composition to a common scale, measured in numbers of adult male equivalents. The following simple method, after Cavendish (1997), can be used; it draws on coefficients directly available from the data or culled from the literature.

Equivalence scale adjustments

There are three main ways in which households vary. The first variation is in the number of people present within the household. To allow for this variation, it is sensible, in the household roster section of the questionnaire, to ask about the length of time that each person named as being associated with the household has spent living in the homestead for the accounting period of the household income data (ie one year, six months, etc). Then each household member can be assigned a time-weighting coefficient equal to the proportion of the year during which the individual was resident at the homestead. Summing these coefficients for each household generates a time-weighted measure of household size. Thus, a migrant labourer who returns a few days per year will contribute very little to household size, but may contribute substantially to income (remittances of gifts and cash).

The second variation relates to the needs of different types of household members – in particular, their nutritional needs. Traditionally, children require less food than adults and women require less food than men. To deal with this, the nutritional weights for different sex and age groups developed by the World Health Organization (WHO) (reported in Collier et al, 1986, and presented in Table 2.9) can be used. Using these nutritional weights, each individual within the household should be allocated a coefficient based on their age and gender. Note that

these coefficients treat males aged 15–59 as the *numeraire*, so their application converts each individual into an adult equivalent unit (*aeu*) (Blundell and Lewbel, 1991; Murthi, 1994). The time-weighted coefficients and the nutritional weights must be multiplied together. Therefore, a 20-year-old female who is around for half the year is 0.88 x 0.5 = 0.44. Summing these numbers over all individuals for each household produces an index of household size measured in time-weighted *aeus*.

Finally, conditional upon these prior adjustments, larger households benefit from economies of scale in consumption and production; as a result, they require less income per individual to achieve a given level of welfare. For example, it takes roughly the same amount of firewood to cook a meal for five as it does a meal for one. We can adjust for these economies of scale by using the weights estimated in Deaton (1982) and presented in Table 2.9.[22] Thus, each household's size (measured in time-weighted *aeus*) is multiplied by an appropriate economy-of-scale coefficient to produce a final measure of household size. For shorthand, this can be denoted as household size in adjusted *aeus*: this figure is used to divide through the crude total income per household

Table 2.9 *Coefficients for adult equivalent scale and economies of scale calculations*

Age	Adult equivalent scale		Household size*	Economies of scale
	Male	Female		
0–2		0.40	0 to 2	1.000
3–4		0.48	2 to 3	0.946
5–6		0.56	3 to 4	0.897
7–8		0.64	4 to 5	0.851
9–10		0.76	5 to 6	0.807
11–12	0.80	0.88	6 to 7	0.778
13–14	1.00	1.00	7 to 8	0.757
15–18	1.20	1.00	8 to 9	0.741
19–59	1.00	0.88	9 to 10	0.729
60+	0.88	0.72	10+	0.719

* Measured in number of time-weighted, adult equivalent units.

measure to produce (this author's) preferred measure of household welfare – namely, total household income per adjusted *aeu*.

This method of adjusting income for differences in household size and composition is not perfect. The coefficients are somewhat arbitrary in that they are derived from Sri Lanka. Equally, it has been argued that the differences in the female and male weightings reflect gender inequalities rather then genuine needs, so that using the WHO nutritional weights underrates women's needs. Even if the weights are correct, it may be inappropriate to assume that indicators of nutritional need can be generalized to need expressed over all commodities. It may also be that economies of scale in household consumption and production exist only for commodities such as food and children's clothing. All of these are real problems, and some could be addressed by adjusting the weights or applying the weights only to a subset of income (eg food). However, in general, the mistakes in the subsequent data analysis made by using such scalings will be considerably less than those made by using either unadjusted household income or simple per capita income (Lanjouw and Ravallion, 1995).

Analysing differences among households

The richness of the data collected in the manner described above lies in the degree to which it can be used to explore the interrelationships between the environment, poverty and inequality in rural areas (Cavendish, 1999c) and the economic differences among households (Cavendish, 1999b; Arnold and Ruiz Pérez, 1998). In this latter case, one important comparison is that made between wealth classes. This is often approached by using total income to subdivide the sample into groups with

the same number of individuals. Quintiles (five equal groups) are a natural way of exploring differentiation, but one can work with quartiles, deciles, etc, depending upon the sample size. There are many other ways to split the data – for example, by the gender of the household head (see Box 2.4 and Figure 2.6), educational status, proximity to resources or markets, and so on (Cavendish, 1999a; Deaton, 1997). Likewise, sub-components of overall total income can be examined in greater detail, where these sub-components are thought to be of particular economic or environmental importance. Examples here might

Note: See Box 2.4 on gender differentiation in resource use.
Photograph by Lini Wollenberg

Figure 2.6 *A woman in Lampung, Indonesia, returning from the forest after having collected damar resin, one of the many internationally traded forest products from the country. Women and children play a key role in the harvesting of forest products*

Box 2.4 Gender differentiation in resource use

Allison Goebel

A key characteristic of the social forest is differentiation along gender, wealth and traditional power lines. Women and men collect and use different forest products, often for different uses (Campbell et al, 1993; Goebel, 1998; Goebel, 1997; Watson et al, 1996). Men, for example, are most likely to hunt forest animals, collect and build with poles and work as carpenters. By contrast, women are primarily involved in collecting thatching grass, in using traditional medicines in the household and in collecting and controlling the use of firewood. These differences show up clearly in the income budget shares of households: in one study, the proportion of household income derived from forest resources ranged from 20 to 42 per cent according to the gender of the household head (Cavendish, 1999a). Furthermore, while female-headed and male-headed households are equally likely to be involved in the planting and care of trees at home sites (Watson et al, 1996; Price and Campbell, 1998), men and women undertake different tasks in these activities, and the locus of decision-making authority varies in the two types of households (Goebel, 1997; Watson et al, 1996). The changing nature of household income is also leading to gender-role switching, with many women and children turning to new activities. For instance, women and youth are increasingly playing a role in carving, which traditionally is an activity of older men. Resource scarcity also drives role switching, best illustrated in the case of fuelwood where in the face of scarcity wealthier households switch from headloading to the use of animal-drawn carts; males thus become the wood collectors (Campbell and Mangono, 1994; Goebel, 1997). Given these dynamics, projects designed to improve availability or management of particular resources will necessarily have a gender dimension.

---

be cash income, fertilizers, livestock, wild foods, and so on.

Emerging results

Using the procedures described in this chapter, empirical results from the work in Shindi, Zimbabwe, are presented in Table 2.10. This contains the household accounts valued, aggregated and booked by income source at a fairly detailed level. Usually, household income accounts are presented by income source at a much higher level, such as by the four categories of cash income, gifts, own-produced goods and environmental income. However, this chapter also includes the detailed income source listing in order to demonstrate what type of activities go where in a full set of accounts; to show the considerable variety of economic and resource-use activities undertaken by so many 'subsistence' rural households; and to suggest how data collection needs to be planned carefully when the goal of the work is a comprehensive picture of household economic activities.

Key results are also briefly reviewed. These are explored in more detail in Cavendish (1999a; 1999b; 1999c). The household accounts show:

• A very considerable contribution is made by environmental income to rural household welfare; this comprises fully 35 per cent of average household total income.

Table 2.10 *Total household income (Z$) per adjusted adult equivalent unit by quintile and by detailed income source, 1993–1994*

	Household quintile					
	Lowest 20%	20–40%	40–60%	60–80%	Top 20%	All households
Crop income	335	930	861	1030	3779	6935
Sales of high-value crops (net of non-labour inputs)	0	117	0	115	1 842	2074
Sales of low-value crops (net of non-labour inputs)	255	456	636	633	1 161	3141
Sales of own-produced vegetables (net of non-labour inputs)	79	346	214	270	716	1624
Sales of exotic fruits	1	11	11	13	60	96
Livestock income*	386	312	576	721	1363	3359
Sales of large livestock	25	0	202	247	178	652
Sales of small livestock	349	259	319	284	1067	2278
Sales of livestock produce	0	0	49	20	41	111
Hire of livestock and equipment	12	53	6	170	76	318
Unskilled labour income	658	696	1292	639	992	4277
Local agricultural labour	574	656	839	639	725	3432
Other local employment	84	40	453	0	267	845
Skilled labour income (teaching)	0	520	0	0	6015	6534
Crafts and small-scale enterprises	120	751	621	1579	3101	6173
Beer-brewing income (net of non-labour inputs)	–72	281	183	317	925	1634
Local crafts	189	460	372	1188	765	2974
Trading goods	0	0	4	12	1411	1426
Hire of goods	4	10	63	62	0	138
Remittances	919	2832	3333	8146	13,175	28,405
From commercial farms	578	1904	1072	2707	8151	14,412
From non-farm employment	341	902	2228	5438	4951	13,860
Other non-Shindi employment	0	26	33	0	74	133
Miscellaneous cash income 0	0	106	82	156	344	
Total cash income (excluding environmental cash income)	**2418**	**6041**	**6789**	**12,197**	**28,580**	**56,026**
Net private gifts/transfers of cash	–121	–202	603	1156	291	1726
Net private gifts/transfers of food or seed	–20	–484	–1	–181	–460	–1145
Net private gifts/transfers of environmental goods	6	2	55	11	3	78
Net private gifts/transfers of other goods	–31	–569	194	–21	1845	1418
Government gifts/transfers of cash	85	37	27	30	290	470
Government gifts/transfers of agricultural inputs	285	457	295	554	526	2118
Total net gifts/transfers	**204**	**-758**	**1174**	**1549**	**2496**	**4664**
Consumption of own-produced goods	5682	7895	8992	10,777	16,766	50,111
Maize (net of non-labour inputs)	3211	5193	6081	7413	10,930	32,828

Table 2.10 *continued*

Other grains (net of non-labour inputs)	1470	960	1021	804	1801	6056
Other staples (net of non-labour inputs)	76	228	296	417	308	1324
Relishes (net of non-labour inputs)	498	714	621	994	1173	3999
Large livestock	0	0	0	0	120	120
Small livestock	410	716	814	1013	2235	5190
Livestock produce	8	65	135	128	198	534
Exotic fruits	8	19	25	8	1	61
Input use of own-produced goods	1289	2130	1867	1765	4092	11,143
Inputs into sold beer	1132	1465	1430	1260	3411	8696
Inputs into consumed beer	18	23	12	24	21	98
Inputs into beer for hiring agricultural labour	0	0	19	44	0	63
Use of own-produced goods as seed	89	171	182	113	160	715
Use of own-produced goods to hire labour	21	20	0	90	93	224
Gifts of own-produced goods to others	29	451	223	234	408	1346
Total own-produced goods	**6971**	**10,025**	**10,859**	**12,541**	**20,858**	**61,254**
Gold panning	1300	1246	2337	2694	4736	12,313
Environmental resource use cash income	708	891	1360	2748	1918	7625
Sales of wild vegetables	9	12	31	8	30	89
Sales of wild fruits	62	70	56	30	153	371
Sales of small wild animals	117	71	1	570	93	853
Sales of large wild animals	34	12	61	112	22	241
Sales of wood	40	43	26	54	0	163
Sales of thatching grass	125	176	174	187	527	1189
Sales of other wild goods	0	35	1	0	73	109
Sales of wine	57	57	108	55	188	465
Sales of large carpentry items	14	20	349	914	324	1620
Sales of small carpentry items	1	24	7	4	0	36
Sales of woven goods	73	49	121	0	0	243
Sales of pottery	10	48	27	0	14	100
Environmentally based local labour income	166	274	398	813	494	2146
Consumption of own-collected wild foods	1334	1668	1919	2183	2453	9557
Consumption of wild vegetables	1020	1370	1567	1650	1999	7606
Consumption of wild fruits	127	129	166	147	206	775
Consumption of large wild animals	26	43	33	190	80	372
Consumption of small wild animals	99	57	86	118	58	419
Consumption of wine	48	54	48	61	94	305
Consumption of other wild foods	14	16	18	16	17	80
Consumption of own-collected firewood	1430	1991	2239	2498	3679	11,836
Cooking and heating	1153	1631	1918	2170	3002	9874
Brick-burning	20	21	56	50	73	220

Brewing-sold beer	256	331	262	259	601	1710
Brewing-consumed beer	0	8	0	5	2	15
Brewing-beer for hiring agricultural labour	0	0	3	14	0	17
Consumption of own-collected wild goods	146	187	203	192	224	952
Wood goods	14	0	0	11	6	30
Large carpentry items	3	2	7	43	0	55
Small carpentry items	21	31	35	20	32	139
Woven goods	30	39	31	35	12	147
Pottery goods	14	9	13	2	10	46
Yard brooms	24	63	62	36	95	280
Toothsticks	23	28	33	29	43	156
Wild medicines	15	13	15	15	16	74
Gifts of environmental goods to others	2	2	7	2	10	23
Use of environmental goods for housing	574	437	1019	975	1471	4476
Wood for wall poles	38	76	64	166	369	713
Wood for cross-poles	21	20	29	27	73	170
Wood for roof beams	81	84	143	111	265	684
Thatching grass for roofing	92	93	112	101	199	597
Thatching grass for roof repairs	181	53	195	174	77	680
Earth for bricks	162	111	476	395	488	1632
Use of environmental goods for fertilizer	130	166	190	150	244	879
Leaf litter	42	30	41	72	65	250
Termitaria	87	136	149	78	171	622
Use of environmental goods to hire labour	0	0	0	0	8	8
Livestock browse/graze of environmental resources	599	1722	1884	2652	3328	10,186
Cattle browse/graze of environmental resources	367	1589	1532	2210	2559	8257
Donkeys browse/graze of environmental resources	83	65	136	179	216	680
Imputed value of ploughing gifts	149	68	216	263	553	1249
Total environmental income	**6220**	**8308**	**11,151**	**14,092**	**18,053**	**57,825**
Total income	**15,813**	**23,616**	**29,974**	**40,380**	**69,986**	**179,769**
Average budget shares						
Cash income (excluding environmental cash income) plus net gifts	16.1	20.3	26.6	34.0	38.9	27.6
Own-produced goods	44.1	42.4	36.4	31.0	32.1	37.2
Total environmental income	39.7	37.3	37.1	35.0	29.0	35.2
Quintile share of total Shindi income	8.8	13.1	16.7	22.5	38.9	100.0

Note: All data are from 1993–1994; US$1 = Z$8.09 in January 1994.

* The fact that livestock income is included here implies double-counting (according to the method used in the section on 'When own-reported values cannot be used: assumptions, omissions and other techniques') because it is already included as a value under livestock browse/graze. However, it reflects an error in the original (Cavendish, 1997).

Source: Cavendish, 1997

- As an income source, environmental income is nearly as important as own-production and much more so than non-environmental cash income.
- Total environmental income is comprised of a wide number of different, sometimes small, individual income sources, highlighting the importance of care in questionnaire design and care in selecting an appropriate recall method if the true contribution of natural resource to rural households is to be uncovered.
- The most important environmental income sources during 1993–1994 were firewood consumption, gold panning, the consumption of wild foods, the value of livestock browse and graze and environmental cash income, although these are likely to

vary from year to year. The quintile analysis demonstrates that environmental income is much more important to poorer households than to richer households.

These data emphasize the huge importance of natural habitats and/or communally held areas to rural households. Conversion or expropriation of these areas would impose a significant welfare loss on rural households, even though this loss would not be measured in conventional household income analyses or the national income accounts. Furthermore, this loss would be felt disproportionately by the rural poor, who are already living at alarmingly low levels of income. The data also suggest that standard measures of rural household incomes and welfare are

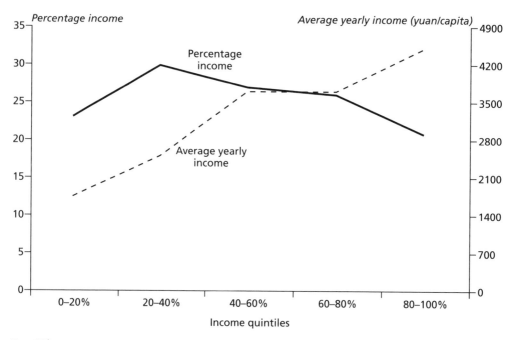

Note: US$1 = 8.5 yuan in December, 1994.
Source: Ruiz Pérez et al, 1998

Figure 2.7 *The relative importance of bamboo production (per cent of income from bamboo) and the total income from bamboo for each income class during 1994–1995 in Anji County, Zhejiang, China*

significantly mis-estimated: in most rural areas attention must be paid to environmental resources in order to gain an accurate understanding of household welfare and household choices.

Another study that has seriously attempted to calculate the contribution of environmental resources to rural household incomes is the much-cited and pioneering study by Jodha (1986). Jodha estimates, based on data from 82 villages in India, that resource uses from the commons comprise between 9 and 26 per cent of the average annual incomes of poor households, and between 1 and 4 per cent of the average annual incomes of rich households.

Figure 2.7 illustrates a quintile analysis for households in Zhejiang, China, for income from bamboo plantations. Across all households, bamboo is the second-most important income source (25 per cent) after off-farm income (43 per cent). Bamboo is a relatively more important income source for the middle-income group than the low- and high-income groups. Absolute income from bamboo increases as we move from the lowest to the highest income quintile.

Conclusions

The main aim of this chapter was to describe a method of integrating a set of unorthodox environmental goods into household income accounts. The chapter detailed how to collect comprehensive data on households' resource use by implementing a locally targeted, random household questionnaire. It also explained the method of booking this data in the household accounts. The chapter discussed at some length the problem of valuing environmental resource uses. With the exception of livestock browse and graze (and some other smaller resources), for which special treatment may be required, it is quite reasonable to use respondents' own estimates of total resource use values, since these household estimates produce aggregate implicit unit values with acceptable properties. In many cases, these values are based on local trading prices; however, even when environmental goods are not widely traded, sensible estimates of resource values can emerge. The chapter also looked at certain goods and service flows that present specific accounting problems, such as beer-brewing and crop incomes, and presented methods for overcoming difficulties in these areas. Finally, it outlined the equivalence scale problem and again discussed the assumptions adopted in constructing a measure of individual welfare (total income per adjusted adult equivalent unit) that allows comparison to be made across households of differing demographic types.

Notes

1   To this end, the questionnaires used for the Shindi study are available from the author.

2   Deaton (1997) provides an excellent, detailed discussion of both the design and content of quantitative household surveys and the econometric issues that arise from the use of survey data. Martin (1995, section 4.3) gives an equivalent description of survey work in the context of ethnobotany.

3   Once-off fortnightly recall is typical of such surveys as the 'Living Standard Measurement Survey' of the World Bank (Grootaert, 1982; Grosh and Muñoz, 1995).

4   For example, in the Shindi work the author drew heavily on Bourdillon's (1987) excellent description of Shona culture and religion.

5   More accurately, the data need to be aggregated to construct some measure of household welfare, which could be income, expenditure or consumption. Following the case study on which this chapter is based, the author calculates household income and classifies resource uses accordingly. A different classification would be required if the researcher chose to work with another household welfare measure. Nonetheless, many of the aggregation principles would remain the same.

6   Usually, at the very least, labour is required to acquire the resources.

7   For example, remittances of food and other in-kind goods from absent household members; non-environmental gifts and transfers from other households; free goods from the government, such as food, seed, veterinary support, school fee payments, etc.

8   For example, crops grown and consumed by the household; livestock and livestock produce reared and consumed by the household; and the use of crops and livestock produce to make other goods, such as grain crops to brew beer.

9   To be strictly accurate, the prices used in valuing inputs and outputs must reflect the *opportunity cost* of the good or service in question. Opportunity cost means the next most valuable use of a good in economic activity. Where markets exist and are undistorted, the price equals the opportunity cost. Where this is not the case, adjustments to prices may be required. See Dinwiddy and Teal (1996) for a fuller explanation.

10   Beer is not just brewed to generate cash. Once a year, many households in Chivi brew beer specially to honour the ancestral spirits that are believed to protect their homesteads and benefit their crops (Cavendish, 1997). This beer is brewed in a special way, in that metal implements cannot be used, the beer is consumed at a particular location, possibly under a tree associated with the ancestor's spirit, and the first cup of the beer is spilt on the earth in order to assuage the ancestor's thirst. Needless to say, valuing this type of beer-brewing is extraordinarily difficult. In other circumstances beer from certain fruit trees, eg *Sclerocarya birrea*, may not be sold according to customary practices. Where these practices are upheld, valuation becomes exceptionally difficult.

11   US$1 = Z$8.09 in January 1994.

12   In this case, crops and firewood were valued at local market prices. Note we use the term 'net' with caution, as the resultant values are not net of labour inputs – see the section on 'The treatment of labour costs'. This holds for all uses of the term 'net' in this and subsequent sections.

13   These values are derived from livestock and, ultimately (at least partly), from wild resources: graze and browse.

14   Because these environmental inputs into agricultural production are not measured in standard surveys, it implies that agricultural incomes and returns are usually overstated in other studies.

15 Establishing a consistent set of quantity measures is often made easier by the relative poverty of the areas studied, which means that each household generally has a small number of purchased goods and that these goods are highly standardized. Thus, one household's 'plate' of fruit would be exactly the same size and make as the next household's, so that goods that are measured in this way (for example, fruits and fish) have clear measures of quantities consumed or sold. Even though households may not be able to measure their resource uses in Western measures such as kilograms and litres, the units of measurement may, nevertheless, be compared across households, which is all one needs. Other common measures in the study of Cavendish (1997) were 90kg maize bags, 20 litre buckets, 1kg and 2kg sugar bags, cups, lids and bundles for foods, and carts, wheelbarrows and bundles for wood, leaf litter, termitaria soil and thatching grass.

16 See Chapter 4 for a more in-depth discussion of using substitutes for resource valuation.

17 In a classic article, Deaton (1988) developed a method to identify the price impact of quality differences from data that do not contain quality measures. The method relies on spatial variation in prices over different regions, and so is only usable with data sets that have been collected in a number of regions.

18 One of these, *Aloe* spp ('*gakekaye*'), was widely used for curing chickens of flu and spots.

19 For example, different studies in Zimbabwe have put the proportion of total cattle values accounted for by draft power at anything between 30 per cent and 48 per cent; by manure output at anything between 3 per cent and 32 per cent; and by milk at anything between 6 per cent and 41 per cent (Scoones and Wilson, 1989).

20 Ten per cent is a common figure assumed for discount rates in developing countries (see also Chapter 5). It was used here, although a higher rate, reflecting private rather than social perspectives, may have been more appropriate (see 'Choosing a discount rate' in Chapter 5). T was set at five years because 90 per cent of the sample households' cattle were beneath this age, suggesting that few cattle made it past this point. Increasing T to six or seven years makes a small difference to the derived grazing value y_0; however, increasing T to ten years would reduce y_0 to a much greater degree and thereby could make a significant difference to the household accounts. So, in setting T, it is sensible for researchers to get a reasonable idea of livestock lifespan in their area of research.

21 This is not true of donkeys, whose meat is, for unexplained reasons, never eaten.

22 Deaton (1982) does not calculate scale factors for households of more than five adult equivalent units. This author follows Collier et al (1986) by assuming that the marginal cost of each *aeu* subsequent to the fifth is constant.

Chapter 3

Understanding local and regional markets for forest products

Michele Veeman

Introduction

This chapter is concerned with the operations and functions of markets for forest products. The focus on markets is on their basic role in the process of exchange and distribution. These are fundamental activities in the organization of all societies (Trager, 1995). The harvesting of wild plants, grasses, fruits, bark and wood for domestic use has traditionally been an important subsistence activity that has provided inputs for household use for many rural people (Martin, 1995; Cunningham, 2000; see also Chapter 1). In rural areas, where many people do not have access to productive land or jobs, poverty is prevalent and harvesting of forest resources can provide the means to barter for needed items. With increasing commercialization of many forest products, the harvesting and sale of these products has become important to the livelihoods of many rural people (Campbell and Byron, 1996; de Beer and McDermott, 1996; see also Boxes 3.1 and 3.2 and Figure 3.1). The markets that have developed for this purpose are the focus of this chapter.

If it is understood how the markets for forest products work, it may be possible to assess particular markets in order to judge how well these operate and whether

they may be improved to the benefit of buyers, sellers and others. In contrast to the markets that operate for processed or consumer products in large urban centres, relatively little is known about the decentralized markets for forest products that operate in many rural areas. Markets for such items as gathered foods, craft products and firewood are often informal in nature and there are few records and little information about these markets (Brigham et al, 1996). Even so, as Box 3.1 shows, they are important to the livelihoods of many rural people.

One indication of the diversity and importance of forest products markets can be seen from the limited data on NTFPs that enter international trade channels. One important non-timber forest-based export has been wild and cultivated rattan supplied to world markets from south-east and east Asia. This has been exported either as poles, cores or peels or as furniture, matting and crafts (Loke et al, 1998). Loke et al note the rapid growth in Indonesia's exports of furniture following a progressive ban on its exports of unprocessed and unconverted poles during 1986–1988. Rattan furniture exports from this nation were reported to have

BOX 3.1 MARKETS FOR NON-TIMBER FOREST PRODUCTS IN THE HUMID FOREST ZONE OF CAMEROON

Ousseynou Ndoye, Manuel Ruiz Pérez and Antoine Eyebe

Local markets play an important role in enabling forest-dependent households to realize a significant part of their cash income through the sale of NTFPs (Ndoye et al, 1998). Increased urbanization associated with migration from rural areas to cities seems to be a significant factor that expands the size of local markets for NTFPs. Four major forest products in the humid forest zone of Cameroon are *Dacryodes edulis* (sold for its fruit), *Irvingia* spp (fruit), *Cola acuminata* (cola nut) and *Ricinodendron heudelotii* (a condiment used with fish). The quantities marketed are significant, valued at US$1.75 million or more during the first half of 1995. There are more than 1000 traders in this system, mainly women. Marketing margins obtained by traders, expressed as a percentage of the value of sales, vary between 16 per cent for *D. edulis* and 30 per cent for *Irvingia* spp. The study confirms the role of these markets as a source of employment and income, not only for gatherers but also for traders. Traders are often accused of exploiting farmers/gatherers. There is, however, a need to recognize that traders carry out many useful marketing functions and that they bear substantial risks from thin and uncertain markets, while carrying out costly distribution functions. The majority (68 per cent) of the traders earn less than the minimum wage from their trading activities, which challenges the conventional belief that portrays all traders as members of wealthy elites who appropriate an unfair share of the profits.

increased from less than US$20 million in 1988 to levels in excess of US$330 million during 1993–1994. Rattan product exports from south China, the Philippines and Malaysia are also significant.

The wide variety of NTFP exports in Indonesia was documented for 1983 and 1987 by de Beer and McDermott (1996). These authors indicate a diverse listing of such exports from Indonesia, headed by rattan (the 1987 rattan export values for Indonesia were reported to be some US$212 million), followed by Illipe nuts (*Shorea* spp) (exports were valued at some US$5 million), Illipe oil (US$3 million) and Jelutong (*Dyera costulata*) (US$2 million). This is followed, in turn, by *Damar oleoresin* (*Dipterocarpus* spp) (US$652,000). While exports of Cassia Vera (*Cinnamomum burmanii*) were also believed to be significant, no data on the value of these were available. The listing by de Beer and McDermott includes materials

for craftwork and furniture, condiments, wild foods, resins for industrial uses, medicinal products, essential oils and fuel. However, much less documentation about the nature of the rural markets for these various products is available.

The paucity of information about most rural markets may be due to their predominantly informal and dispersed nature. In addition, many participants in rural markets do not hold appreciable economic or political power, so that little attention may be paid to their interests by policy-makers and bureaucrats. Uncertainty about the legal status of some resource collection activities from state-regulated or communal forests may also contribute to the lack of documentation of most markets for forest products. However, knowledge of how these markets work, and whether or how they can be improved, is of importance for many reasons. In

---

BOX 3.2 MARKETS FOR SAVANNA PRODUCTS IN SOUTHERN AFRICA

Allison Goebel

In southern Africa, widespread markets for forest products are shaped by the local ecological environment, seasonality and labour regimes (Brigham et al, 1996; Hot Springs Working Group, 1995; Campbell et al, 1997b; Campbell et al, 1991; Graham et al, 1997; Goebel et al, 2000). Seasonal markets exist for indigenous fruits, with households realizing major short-term incomes from the sales. In the Hot Springs area, significant increases in household income can be achieved through the selling of quelea birds, although this benefit varies dramatically from year to year. Many households also market caterpillars. Honey can be a significant earner for some households. Other commonly sold woodland-derived products include benches, chairs, stools, hoe handles, axe-handles, adze handles, yokes, mats, baskets, thatching grass, bricks (which require firewood in preparation), mice, termites and traditional medicines.

For a few households, the marketing of woodland-derived products has outstripped the importance of agricultural activities as an income source. This is true for some participants in the expanding craft industry along the major tourist routes of Zimbabwe (Braedt and Standa-Gunda, 2000). In the Hot Springs area, for example, some people have reduced their agricultural activities due to extremely dry conditions and focus instead on the production of crafts from baobab bark. Gross per annum incomes as high as approximately US$2000 per household have been recorded from such activities for the larger trading stores (Kwaramba, 1995). Along the Beitbridge–Masvingo and the Bulawayo–Victoria Falls roads, huge increases in the numbers of markets for woodcarvings have been observed in recent years (Matose et al, 1997). In the former area, carvers harvest valuable hardwoods as a 'free' resource from the woodlands in communal areas, or purchase inputs from better-wooded resettlement and small-scale farming areas. Fuelling the growth in these markets is the large increase in foreign tourists, particularly coming from South Africa (see Figure 1.4). This is at least partly linked to structural adjustment, which over the last five years has caused significant devaluation of the Zimbabwe dollar vis-à-vis the South African rand. Many buyers are dealers who evidently are able to realize profits through the resale of carvings in South Africa and abroad. In general, the extreme dynamism of conditions and livelihoods in smallholder systems needs to be recognized (Campbell et al, 1997a).

---

particular, achievement of better market performance is significant to rural people for whom these markets can be a means of providing or improving household incomes and livelihoods. Knowledge of how markets for forest products work may also be helpful in developing better policies to help manage the underlying resources, particularly where there are concerns about the ecological sustainability of the resource, relative to current or prospective harvesting or extraction.

The following section turns to an overview of basic concepts of markets and market operations. In the section on 'Some features of rural markets', general features of rural markets are outlined. This is followed by a discussion of economic perspectives of market performance and an overview of market research tools that may be applied to obtain a better understanding of how markets work.

Photograph by Reidar Persson

Figure 3.1 *In the dry forests and woodlands of Africa, firewood is often valued relatively highly. Here a farmer in Niger transports firewood from the forest to the market using a donkey. The money from firewood sales makes a significant contribution to his household's cash income*

Some basic concepts of resource markets

What are the basic features that describe a market?

The basic elements of a market (as perceived by economists) focus on the process of exchange, which involves the existence of buyers and sellers and the means of communication that enable them to exchange the goods or services in question. Even so, there are many ways to describe markets and some of these can be the basis of useful forms of classification. Some market descriptions are based on different types of market institutions through which exchange may be organized. Another way of describing markets relates to different levels or functions of a marketing chain. This section notes the different ways of describing markets and outlines other features of exchange and distribution.

Means of price discovery

One way of describing specialized markets is to summarize how they are organized in

69

terms of providing for exchange, as with the process of price discovery. For example, auction markets or private treaty markets describe two different ways in which price discovery may be pursued in organized market exchanges. Raw wool and cured leaf tobacco are two commodities that are frequently sold by auction through organized exchanges at the wholesale level of the marketing chains.[1] For other commodities, prices may be established through the process of rapid public statements ('public outcry') of bids and offers in an organized exchange. This has been the traditional means of trading on both stock market and commodity futures exchanges, although the process is now being replaced in many such exchanges by computer matching of bids and offers. Private treaty trading, where the buyer and seller negotiate the price and other terms of exchange is another frequently used basis by which goods are exchanged. This occurs in a wide variety of institutional settings. Private treaty sales are typical of traditional and other informal markets in rural areas.

Levels in the marketing chain

Markets can be very usefully described by the level of a marketing chain, providing a summary of the particular functions of the chain in question. Examples of this type of market description are retail, wholesale and 'farm-gate' or 'wood-lot gate' local markets. Several different stages of wholesale markets may be involved in a marketing system as products move through a marketing chain from producers to consumers. Thus, forest products may move initially from producers to small traders who, in turn, may sell to bulking-up wholesalers, as product is accumulated into the larger lots that are traded in bulk-sale whole-

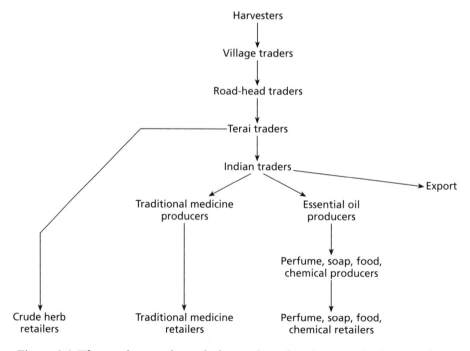

Figure 3.2 *The marketing channels for medicinal and essential oil plants from Nepal to India*

BOX 3.3 THE MARKETING OF MEDICINAL AND ESSENTIAL OIL PLANTS FROM NEPAL

The marketing of NTFPs in Nepal illustrates some of the complexities of market channels (see Figure 3.2; Edwards, 1996). There is a network of village traders and marketing agents who have valuable knowledge about local medicinal and oil sources and a close relationship with harvesters. The close relationship, embedded in the Buddhist cultures of the region, is characterized by tight-knit communities with long-term forms of recipirocal cooperation. Their participation in the marketing process decentralizes and improves the efficiency of the initial stages in the marketing of products to the point where 'road-head traders' are involved. The village-level traders either work as independent traders or are agents for the road-head traders. Some harvesters sell their products directly to road-head traders. While this gives them greater economic returns, they must have access to market information (in order to know when to benefit from a journey to the road-head). They must also have sufficient household labour to take care of the household's farming operations during the journey and the financial capacity to accept risks associated with storage losses and unforeseen price changes.

Road-heads are the main centres of activity in NTFP trade within Nepal. These are the points on the road system where the product is loaded onto lorries for transport to India. Road-head traders amass sufficient stock (approximately 5 tonnes) prior to moving the material in a lorry to the district headquarters, where royalties can be paid and permits received for export to India. The road-head traders in the north of the country in turn sell their products to the Terai traders visiting from the south.

Terai traders live in towns near the main east–west highway. They manage their operations over large territories that include several road-heads. The Terai traders are typically Marwari, a culture group with close links to India. They are in regular telephone contact with Indian traders to obtain the latest market information, and access to this information is carefully controlled. These traders tend to operate as a cartel, preventing would-be new entrants from breaking into the trade structures and reducing the bargaining power of road-head traders, which may reduce profits to harvesters and village traders.

The traded goods then move to Indian traders in the large cities of India. At each large centre, there are wholesalers and commission agents who buy from the Terai traders. These traders in turn sell to the medicine and essential oil industries or export the materials. The products then are sold to the retailers.

---

sale markets. Subsequently, such products may be traded in smaller lots in bulk-breaking centres, where buyers may include small retailers or large consumers (see Box 3.3).

Geographical concepts of markets

Markets can also be described as being centralized (as with some of the examples of specialized exchanges or large urban markets) or decentralized (typically the case in rural areas where numbers of small local markets may be found within the regional area). Economic geographers have identified spatially based hierarchies of markets that perform rather different functional roles (see 'Some features of rural markets') (Skinner, 1964, 1965; see also Cunningham, 2000). This includes:

- central markets: the large markets where wholesaling occurs and numbers of specialized markets may be found;

- intermediate markets, which operate one step closer to standard markets;
- standard markets, which are considered the effective interface for the end sale of city-based or import goods and the entry point for rural-based goods to move into larger regional markets; and
- minor markets, which are locations for local exchange by local people.

Similarly, markets may be described by economists on a geographical basis according to the scope of the forces of supply and demand – for example, when referring to the world market for lumber, which is a description that reflects economic linkages between numbers of large markets. Even so, many local markets also exist for various wood products, and the localized influences of supply and demand are particularly evident at this market level.

The basic elements of market exchange

In simple terms, the basic elements of market exchange are the existence of buyers and sellers, with the means of communication that enables them to make an exchange of the good in question. Buyers and sellers must have sufficient information and a means of exchanging information (communicating) so that they can agree on a price and any other necessary terms of exchange. Other terms of exchange to be determined in any market situation may include delivery arrangements, financing arrangements and quality specifications, including a guarantee of the quality of the good if it is bought on the sight-unseen basis of a grade description.

The means of communication required for markets to operate need not involve face-to-face communication between buyers and sellers, but can involve long-distance communication, either directly (between individual buyers or sellers) or indirectly (through the mechanism of a marketing agent, who sells on a commission basis), or even through a clearing house (as with organized commodity exchanges). Consequently, the exchange of goods need not be directly conducted between the buyer and seller, but can involve marketing institutions or agents who perform a number of the marketing functions that are discussed in the following section. Specialized trading arrangements, as with sale through agents, requires a degree of trust between market participants and the existence of a social system that ensures the enforcement of contracts. In traditional rural markets, the sale of resource and other products in local markets typically involves face-to-face exchange between buyers and sellers.

The process and functions of marketing

To understand how markets work, it is useful to examine features of the process of marketing primary products, such as resource and farm products. Marketing is the performance of all the activities necessary to transform a raw product from its point of production, harvest, gathering or collection, to the point of final consumption, whether as a food, medicinal or household item. It is convenient to look at these activities in terms of three sets of marketing functions. These are the physical functions required for distribution, the functions that are directly associated with exchange, and the facilitating functions of marketing that are necessary for exchange and distribution functions to occur.

Knowledge of marketing functions is helpful in understanding how markets work. The functions that are involved in marketing encompass inputs of one form

Table 3.1 *The functions of marketing*

Functions	Activities
I Physical functions	• Transportation • Storage/holding inventories • Processing
II Exchange functions	• Buying (this may include assembly of products from different sources/areas/quality assessments) • Selling (this may include product planning/development/promotion) • Price discovery/establishment
III Facilitating functions	• Information provision • Financing • Risk-bearing • Grading/standardization • Good governance (security of people and property; provision of civil law that gives the basis for contract enforcement/ bonding of agents; provision of other public institutions that enable markets to operate; regulatory activities that accommodate market failures such as externalities or public goods)

or another, whether time, effort or money. Consequently, the functions of marketing contribute to the costs of marketing, and also to the value created by marketing and the effectiveness and efficiency of particular markets. The functions of marketing are listed in Table 3.1.

Physical functions

The physical functions of marketing that are required for the distribution of goods involve *transportation* from the point of initial production or harvest to the ultimate consumer. For primary or resource commodities, such as the products of farms and forests, the raw product is typically produced or harvested over geographically distributed areas that generally are a considerable distance from the central markets associated with major consumption centres. Transportation costs can be a relatively high component of total marketing costs when there is: regional dispersion of production/harvest; low value of product, relative to its bulk; poor transportation facilities and/or perishable

products. The existence of markets for various rattans, which are second only to timber as the most important forest product in much of south-east Asia, is largely dependent upon demand for cane furniture in Europe (Dransfield and Manokaran, 1994), and thus largely dependent upon transportation to access European furniture markets.

Lack of transportation infrastructure will be reflected in higher levels of transportation costs in the process of marketing, and these costs are normally reflected in the prices received by harvesters or gatherers. This was demonstrated, for example, in a study of the marketing of vegetable ivory from tagua palm (*Phytelephas aequatorialis*) in Ecuador. In Esmeraldas Province, which is wetter and has a poorer road network than in other study districts where tagua is harvested, Southgate et al (1996) found that that transportation costs were appreciably higher; households that were able to deliver directly to market intermediaries were paid prices that were 22 per cent higher than if delivery was made at the farm gate. Even so, relatively costly rapid transportation may enable markets for high-valued forest products to be established and maintained. Cunningham (2000) points out the importance of the role of air transportation to transfer young leaves of the African medicinal plant khat (*Catha edulis*). The desired quality of khat leaves for chewing deteriorates rapidly with time. This product is sufficiently valuable that it is cultivated and a high level of organization is applied in order to transport it rapidly, by air, from farms in Kenya to markets in Somalia or even further afield in Europe.

Like transportation, the function of *storage*, involving the holding of inventories, is likely to be necessary during several stages of marketing; this is so that products are available at the time and place desired by consumers (see Figure 3.3). Storage may be a major component of marketing costs when forest products are only seasonally available for harvest or when consumption has an appreciable seasonal component, reflecting seasonal preferences for consumption (preferences and demand increase at particular times of the year, such as during Christmas or other seasonal celebrations). In the historic export trade of gum arabic from the Sudan, the lack of safe storage between seasons for both the bartered good, *guinée* (blue cloth from India), and gum arabic led to wild swings in price that were a major market disruption (Webb, 1985).

A more recent example of the importance of storage as a marketing function is seen in the case study of woodcarvings in roadside markets in southern Zimbabwe (Box 3.4). Sales of woodcarvings to travellers can be dependent upon holding relatively large stocks of carvings to display to potential buyers; however, costs

Photograph by Manuel Ruiz Pérez

Figure 3.3 *A stall for the sale of medicinal products in Mbare market, Harare, Zimbabwe. The storage function of the market is amply illustrated*

are involved in maintaining such unsold stocks, including the risk of theft, loss or deterioration and the opportunity cost of the stocks. An example of deterioration of a forest product during the period when the product is displayed for sale is reported by Zimbabwe weavers of craft products made from the bark of baobab trees (*Adansonia digitata*) (see Box 7.2 in Chapter 7). Rural households in the regional area north of Birchenough Bridge collect and weave bark into craft items, such as woven mats, hats and bags, which are often sold to travellers from simple uncovered roadside stalls (Veeman, M et al, 2001). In interviews with bark-craft weavers and vendors in this region, sellers noted that, over time, exposure to weather of items that are displayed at the roadside fades the natural dyes used by most weavers. Weathered baobab products typically have to be sold at a lower price than unweathered craft products sold by craft merchants from enclosed stores in this region. Added costs of sheltered storage would have to be incurred by householders to avoid weathering. Consequently, many householders do not display their full stock but will bring these from their nearby homestead if requested.

Storage is one means of overcoming seasonality if the product is storable. However, storage may be particularly difficult or costly if a product is perishable. Together, transportation and storage are two important components of the physical functions of marketing that are necessary for distribution. The performance of these two functions can involve appreciable marketing costs, while adding considerably to the value of forest products for the final user.

Processing is the other major 'physical' function, reflecting the need for physical transformation of many forest products prior to their purchase and use by consumers. For many forest products, transformation of the raw product into the form that is desired by, and of value to, the consumer is a major marketing function that adds significantly to the value of a finished product. Consequently, processing is likely to be a major component of the marketing costs of the final product. One example of the contribution of processing to market value and costs is the transformation of the raw agricultural products of livestock or grain into the foods of meat, flour or bread. Another example is the transformation of raw forest products, such as wood, into firewood or charcoal or the transformation of canes, wood, bark, grasses or reeds into artisan, craft or artistic products, such as furniture, baskets, mats or carvings. In the agricultural and fuelwood examples, the processed products are likely to be relatively standardized, and thus costs are also likely to be fairly standard. In contrast, the value added from transformation of resource inputs into some craft or artistic items can be highly variable, reflecting the technical and artistic skills of the artisan and the production process.

Exchange functions

Exchange functions involve buying, selling and price establishment. These will occur at the various stages of a marketing channel and in different institutional structures. In a market setting, prices arise from a process of price discovery, and price levels reflect the nature of supply (associated with the availability of the good and sellers' willingness to provide this at particular prices) and the nature of demand (associated with the preferences of consumers and their willingness to purchase at particular prices). In a market context, economists conceptualize supply and demand as functions that relate price levels to the quantities provided for sale (in the case of supply functions) and the

BOX 3.4 WOODCARVING MARKETS IN SOUTHERN ZIMBABWE

Ivan Bond

Recently in Zimbabwe, informal craft markets have proliferated in urban areas and along the major tourist routes in rural areas. This increase has been driven largely by the changes to the economy as a result of the economic structural adjustment programme (ESAP) and growing demand for craft products from increasing numbers of regional and international tourists. A wide range of crafts manufactured from stone, clay, wood, wire, steel and fibre are often sold in these markets. The increasing use of wood has serious implications for the management of indigenous woodlands in Zimbabwe.

Research methods

A general survey of all craft markets along the Masvingo–Beitbridge Road was conducted and two monitoring systems were used to collect site-specific data (Braedt and Standa-Gunda, 2000). Between June 1997 and July 1998 (inclusive), all purchases of materials (including wood and tools), hours worked and sales by 17 randomly chosen carvers were monitored. The net financial benefit per unit hour to the carver was calculated. To monitor the market systems, three markets were selected in which the number and roles (carver, retailer or both) of participants were recorded and focus was on the sale of carvings. All the carvings in these markets were measured (length, width and height), and tagged to allow enumerators to record the actual sale price and the date when the carving was sold (see Table 3.6). The methodology was applied to three markets over a time span of between 18 and 22 months in 1996, 1997 and 1998. During this period, a total of 2774 carvings were tagged, of which 1093, or 40 per cent, were sold. In July 1997, US$1 = Z$11.5, while in July 1998, US$1 = Z$18.

Results

Market structure

The Masvingo–Beitbridge Road is the major tourist corridor for South Africa visitors to Zimbabwe. In February 1997, the general survey visited 26 recognized craft markets along the road, of which 21 had started in the previous five years. The markets tended to be located in those sections of the road that passed through land that was under smallholder agriculture. The type of products sold in the markets has changed significantly over the last 20 years. Before 1980, there was one market offering primarily clay and some needlework products. In 1984, stone-carving first entered the markets. By 1997, however, the dominant product was woodcarvings. Based on the sample market survey covering three markets, the most common species of wood used for carving in all three markets was *Afzelia quanzensis*. Of the 2774 carvings tagged, 68 per cent were *A. quanzensis*. The total volume of the tagged carvings was 161 cubic metres (m^3) and of this 45 per cent was *A. quanzensis*. Other key species, by volume, used by the carvers were *Combretum imberbe* (26 per cent), *Kirkia acuminata* (8 per cent) and *Pterocarpus angolensis* (6 per cent). There was a close relationship between the species used and the type of product carved. For example, 44 per cent of the *A. quanzensis* carvings were hippopotamuses, 72 per cent of the heads were carved out of *C. imberbe* and 60 per cent of the *K. acuminata* carvings were giraffes.

Both the de jure and de facto rules governing the operation of the market were assessed. Under the Communal Lands Forests Products Act (1984), harvesting indigenous timber for commercial carving is illegal. However, the legislation is seldom enforced. Local 'traditional' rules about the use of the woodland are loosely enforced. The result is that indigenous woodlands can be considered an open access resource. The operation of the markets themselves falls under the rural district councils (RDCs). However, of the 26 markets enumerated only 5 were sanctioned by the appropriate RDC. Within the markets there is generally a committee, chaired by the founder of the market. Participants pay a joining fee of between Z$10 and Z$75 in order to sell their carvings at the market. It is unclear what the fees were used for. Within the markets, the behaviour of sellers is regulated by a loose set of rules. However, these are often loosely interpreted and vary with the demand for products. Serious infringements (eg theft) were punished by expulsion.

Conduct
The degree of market integration between the three markets was analysed using unit root and cointegration tests. The results suggest a high degree of price integration between the three surveyed markets, which was probably due to their close proximity (the distance between the first and third market was 4.8km), the multiple functions fulfilled by market participants (carvers and sellers) and the practice of selling carvings on behalf of absent sellers. The actual price received for each carving was, on average, 73 per cent of the expected price. Analysis showed that size was the only carving characteristic contributing significantly to price. Tree species and product type showed no significant impact on price. Most tourists are not aware of, or do not care about, the tree species being traded.

Performance
A simple gross-margin analysis, a calculation of the value added to indigenous timber species and a financial model were used to analyse the performance of the woodcarving markets. This showed that for a sub-sample of comparable carvings (hippopotamus carved from *A. quanzensis*) the net return* per unit hour of carvers' labour was approximately Z$10. This was approximately double the government gazetted minimum wage rate (Z$4.04 per hour) and the estimated opportunity cost of local labour (Agricultural Labour Bureau, pers comm; Campbell et al, 2000a). However, of the 2774 carvings tagged, only 1093 (40 per cent) were sold. For those carvings that were sold, the median time in the market before sale was 11 days.

Discussion: the future of the woodcarving industry?
Overall, woodcarving is an important economic activity in an area characterized by low and unreliable rainfall. Mechanisms need to be investigated which can legitimize woodcarving as an economic activity, as well as contribute to the improved management of woodlands in the area. The study suggests that current profit margins are very narrow and the resource is heavily depleted. Consequently, alternatives – such as incentives for carvers to use other species, particularly exotics – need to be considered.

* This excluded the opportunity cost incurred because of delays in the sale of carvings.

quantities that will be purchased (in the case of demand functions), at alternate prices, while holding constant the other influences that may affect supply or demand. An example of 'another influence' that could shift the supply function is a change in the numbers of harvesters because of a change in regulations or road accessibility. For demand, 'other influences' or demand 'shifters' could be a change in the number of consumers, due to a change in tourist visits (Figure 1.4 in Chapter 1) or a change in the income levels of consumers.

The supply function for a good is derived from producer-level decisions and behaviour. It reflects a positive relationship between sellers' willingness to sell and the level of price. This is because with higher levels of prices, all else being equal, harvesters and producers will expend more effort to collect or process goods to sell. However, the demand function for a good develops from consumers' preferences and constraints and reflects an inverse relation between price and quantity demanded (all else unchanged), since at higher prices buyers will seek other sources or products and thus will postpone or reduce their purchases. At 'downstream levels' of the marketing chain (ie at wholesale and producer levels of the marketing system), the concept of derived demand applies, since demand functions that apply at such levels are derived from the primary (consumer-level) demand function. Derived demand functions reflect the costs of processing and other marketing functions, as well as the primary (consumer-level) demand function.

The process of price discovery, which results from the concepts of supply and demand, is a fundamental feature of markets and the process of marketing. As noted earlier, this may occur at different levels of the marketing chain and be organized through a variety of institutional arrangements. For instance, this includes auctions at livestock assembly yards, organized commodity exchange operations of open outcry or electronic bidding, and various forms of private treaty trading whereby buyers and sellers negotiate prices after physical inspection of the traded item. The latter process generally applies for most forest products.

Facilitating functions

Facilitating functions of marketing contribute to the effectiveness of how markets work to provide for exchange and distribution, as well as the efficiency of marketing processes. This group of functions includes market information, financing and risk-bearing. Grading and standardization may also be important. Provision of a social environment that leads to enforceable contracts and the absence of theft or graft could be termed 'good governance', and this is another important market facilitation function which is listed, with some general examples, in Table 3.1. Some of the facilitating marketing functions may be provided by specialized marketing agents who operate at various levels of a marketing channel. In other market situations, buyers or sellers may largely provide for some of these functions. Lack of facilitation of markets and marketing processes may reduce the productivity of market participants, reduce the utility and value created by marketing processes and add unnecessarily to the uncertainties and other costs of marketing forest products. Relatively few facilitating services of marketing are available for forest products in rural and regional markets.

Market information is basic to the operation of markets. Accurate market information reduces risk to market participants and allows better planning of harvesting and marketing decisions. Knowledge about prices in other markets is also necessary for arbitrage to occur (trading that occurs in response to the existence of different prices in different markets). Arbitrage enables products to be transported to locations where supplies may be scarce, from regions where supplies may be abundant or where demand has increased. Provision of market information that is accurate and timely can be carried out by private marketing agents; but market information can also have public good features, indicating a potential need for government involvement, as with the need to prevent fraud through accuracy of weights, other measures or quality standards.[2] Differential access to information can be a major source of market power, as shown in the relationship between different categories of traders for medicinal plants and essential oils as these forest products move from remote hillsides in Nepal to the large urban markets in India (see Box 3.3). The differential access of Terai traders to market information stems from access to communications and cultural links with Indian traders who are the source of market information. Under these conditions, relatively lower prices and margins are expected for harvesters and village-level buyers.

An example of the contributions of new information technologies to small traders and rural people is the use of telecommunications in rural Costa Rica, enabling small coffee growers to obtain current market price information from centrally located cooperatives, since these have links, through computers, to sources of information on national and international coffee prices (World Bank, 1999). Another example is the use of cellular phones by farmers in Côte d'Ivoire to get international cocoa price quotations directly from Abidjan. A third example arises from the introduction of the telephone service to several rural towns and villages in Sri Lanka; this has allowed small farmers to obtain first-hand, up-to-date information on wholesale and retail prices of fruits and vegetables in Columbo, the capital city. Since then, farmers have been able to sell their crops at prices that are reported to be 80–90 per cent of Columbo prices – an appreciable increase from only 50–60 per cent of the capital city prices they used to receive (World Bank, 1999). Help from a non-governmental organization called Peoplink has enabled women in Panama to post pictures of their handicrafts on the worldwide web, enabling them to gain access to a world market (World Bank, 1999).

The marketing function of *financing* is necessitated by the costs of holding inventories of unsold products or inputs during periods of transportation, storage and processing. These costs will be greater for products for which there are considerable time lags in processing, transportation or storage, or where there is seasonality in supply or consumption. Where time lags occur, the possibility of risks also increases. Physical or financial risks add to the costs of *risk-bearing*. Risks may arise because of thefts, product spoilage and deterioration, or because unanticipated changes occur over time in prices received and costs paid. Risk-bearing costs may be high for valuable products and where functions of 'good governance' are not well provided, or when there are uncertain or unstable market prices. All these characteristics may be related.

The very high risk level in the terms of exchange for gum arabic, which was the most important trade good exported from Mauritania and Senegal during the late 17th century until the 1870s (when it was eclipsed by the export of peanuts), was reflected in very high returns for successful traders. However, many traders were unsuccessful due to the risks from uncer-

---

BOX 3.5 MARKETING OF *UÑA DE GATO* IN PERU

Wil De Jong

Uncaria tomentosa and *U. guianensis*, known as *uña de gato*, are native to several South American countries. The two species have been used as medicinal plants; according to local beliefs, the plant is recognized as a powerful medicine for a number of diseases (de Jong et al, 1999). More recently, compounds derived from these plants have been recognized by Western medicine as having positive anti-inflammatory effects that are useful for treating numerous diseases. The massive rise in export sales, and the expansion of trade to many countries, is shown in Table 3.2 for Peruvian exports. While the trade has made significant gains for traders and harvesters, there are questions as to whether it is ecologically sustainable.

Table 3.2 Uña de gato *export sales from Peru*

Year	Weight (kg)	Numbers of countries to which species are exported
1993	200	1
1994	20,743	8
1995	726,684	24
1996	346,903	26

In 1995, out of approximately 100 permits that were issued by the government for the export of the product, only 12 companies sold the product packed, ready for sale. The majority of sales comprised unprocessed bark or ground bark. During that year it was rumoured that the government was proposing to ban the export of unprocessed material, in order to capture greater value-added in the country. This is said to have led to the export of greater quantities during that year, so that importers could stockpile the material. In order for greatly increased value-add to occur, it is necessary to increase the capacity of the Peruvian industry to process large quantities. Strict quality-control measures must also be applied and sufficient capital must be invested in marketing and advertising. If the ban goes to the extreme of only allowing the export of completely finished material (eg capsules), it is likely that the Peruvian industry will suffer. It is unlikely that such finished materials will be readily accepted in international markets because of the lack of quality assurance for foreign buyers of the finished product.

Several Peruvian manufacturers and traders have proposed producing their own raw materials through production forestry. This would have significant negative effects on the livelihoods of the small-scale collectors; but a move to vertically integrated production forestry would parallel trends that have occurred with many forest products.

---

tainty in the supply of this forest product and the risks inherent in the availability of *guinée*, for which gum arabic was exchanged prior to its shipment to European markets. During that time, gum arabic was widely used in making paint, paper and glue, in sizing cloth and in a variety of other uses (Webb, 1985).

The rapid deterioration in quality after harvest of the African medicinal plant khat

(*Catha edulis*) is a source of risk for growers – rapid transportation to market is necessary for quality to be maintained to point of sale. If there is unpredictability in government actions, this can be a source of market instability that adds to the risk faced by market participants. For instance, uncertainty about the proposed implementation of laws to restrict the export of raw materials affected the market for the medicinal plant *uña de gato* from Peru (see Box 3.5)

Grading and standardization can be an important marketing function for some forest products. If products are sold on a 'sight unseen' basis, so that the costs of physical assembly for the purposes of inspection are avoided, costs of marketing can be appreciably reduced. Grading and standardization can be achieved if the product has readily measurable characteristics that are relevant to users or consumers, and if the distribution of these characteristics is such that products can be categorized into well-defined groupings of fairly homogenous product, based on relevant characteristics. In practice, grading systems are important for most agricultural or food products that arise from the commercial farming sector and/or where grade standards cannot be readily assessed by buyers, leading to the possibility of cheating or fraud. The ability to apply and enforce grade standards can reflect features of 'good governance'.

Some examples of grading and standardization systems have been applied and maintained by government; for others, this has been pursued successfully by associations of producers and traders. One agricultural example of a successful focus on quality improvement for a food product was the programme run by the national Dairy Development Board in India during the 1970s to improve quality of milk. Increasing demand for milk since the 1950s led many vendors to water milk, leading to a drop in quality overall (quality could not be detected by buyers and honest sellers could not match the prices charged for watered milk). In order to improve quality, the board established village milk cooperatives and provided each cooperative, as well as distributors and marketing agents, with a simple mechanism to measure butterfat content. Consequently, prices could be paid for unwatered milk that reflected this aspect of quality. This not only strengthened incentives to market milk of high quality, but helped cooperatives to improve management and construct processing facilities, as well as establish refrigerated transport. All of this has been credited with major improvements in milk quality and the achievement of higher incomes by producers. Within a decade, producers in the target areas of India were reported to have doubled their incomes from milk sales (World Bank, 1999).

Current lack of buyers' quality assurance, where this is not obvious from inspection of the finished product, may reduce the benefits to local producers from the sale of some forest products, such as medicinal products. An example is seen for the medicinal *uña de gato* from Peru; the inability to guarantee ingredient quality has been suggested as a factor that may hamper the ability to successfully expand Peruvian processing of this medicinal (Box 3.5). Quality improvement can achieve increased returns from forest resources. For example, orienting the processing of rattan cane furniture to high levels of design and quality standards in order to attract the 'high end' of the European market for rattan furniture was a market development strategy advocated for Malaysia by Loke et al (1998). This recommendation was framed in the light of decreasing rattan supplies in Malaysia. Concurrently, increasing supplies of competing cane and bamboo furniture

---

BOX 3.6 RULES GOVERNING THE EXTRACTION OF FOREST PRODUCTS FROM THE VILLAGE FOREST RESERVES OF SAN RAFAEL, LORETO, PERU

- It is prohibited for individuals, families and groups to extract timber from the forest reserve. The extraction of timber is only permitted at the communal level, when the village needs money for communal purposes, such as a new school or medicine.
- NTFPs can be extracted by an individual, family or group.
- Poles and other local construction materials can be extracted by any member of the village. Extraction for commercial purposes is limited to small quantities and to certain periods of the year, such as when rice is harvested.
- All villagers, and individuals from neighbouring villages, are permitted to extract fruits and medicinal plants, either for own consumption or sale. In collecting fruits, leaves, flowers, bark, resins, roots and branches, cutting trees is prohibited. The extraction of particular valuable species is regulated by special rules.
- Timber extraction by non-members of the village requires a permit issued by the village.
- Using the area for agriculture is not permitted.

Note: These particular rules were generated through intervillage discussions.
Source: adapted from Pinedo-Vasquez et al, 1990, 1992

---

from 'lower wage' Asian nations captured the 'lower end' of the cane furniture market in Europe. An example of quality differentiation in craft sales may also be observed in some villages in the island of Bali (Indonesia), where woodcarving is a long-established artisan activity. In the retail sales of carvings, the pricing structure for these crafts may be differentiated by the skill of the carver: items made by master carvers are more expensive than those carved by apprentices.

Provision of '*good governance*' is necessary for the effective operation of markets, among many other features of a well-functioning society. Security of people and property, as well as the maintenance of well-defined and accepted systems of civil law or customary practices, are necessary in order to enforce valid contracts and reduce physical and market risks. Rules governing market operations need not only be those defined by formal institutions, such as government, but may include informal rules and regulations or norms of

behaviour, as established by traditional leaders or informal groupings of users (see Box 3.6). Providing basic institutional infrastructure that allows goods to be moved through the marketing system in order to reach consumers is also necessary for markets to operate effectively, as is providing the means for buyers and sellers to communicate.

Situations of market failure can arise when there is a monopoly or monopsony in the marketing chain (see the proceeding section). Market failure can also arise from the unpriced effects of harvest or production that impose external costs or benefits of consumption on others (externalities). In extreme cases, externalities may involve public goods. Market failure may also occur when knowledge is imperfectly available – for example, when there is asymmetry of information between buyers and sellers. In these circumstances, market prices do not reflect social values and markets do not work well to achieve efficient outcomes. Effective public policy to rectify these effects will be necessary for efficient market operations. Good governance reflects a political, social and economic environment within which public policy can be framed and directed at externalities and other sources of market failure. Good governance also reflects an environment that does not encourage opportunistic behaviour (see below); this may limit effective markets and effective marketing processes. While good governance should be the objective of government policies, the policies and actions of government may have undesirable consequences. This has been the case for the Indonesian sandalwood industry, where policies regulating resource use and promoting industry development have led to the collapse of the resource – in large measure because insufficient attention was

paid to community rights and the impacts of the policies on individuals and communities (Box 3.7).

Opportunistic behaviour, defined as acting in one's own self-interest with guile (Schaffer et al, 1987), involves taking advantage of circumstances in ways that may be defined as corrupt in some cultures. This results in a lack of trust in market participants. Opportunistic behaviour may reflect asymmetry in information or a lack of institutions that can provide information. It benefits individuals at the expense of others in the marketing system, increasing marketing costs and restricting the use of markets. For example, opportunistic behaviour in which grade descriptions or measures cannot be relied upon (resulting in all transactions requiring visual inspection) increases marketing

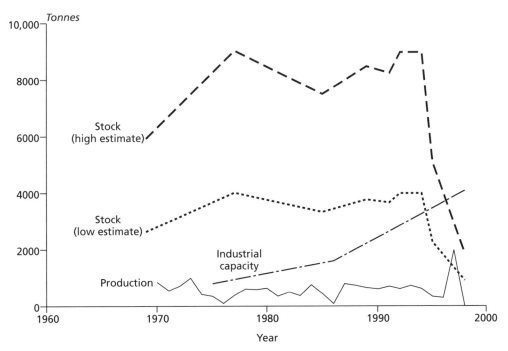

Note: Fitted lines have been added to the three variables that do not have annual data points.

Figure 3.4 *Trends in production, standing stock (with a low and a high estimate) and industrial capacity for sandalwood (capacity is currently met through supply from illegal sources)*

BOX 3.7 CAN SANDALWOOD IN EAST NUSA TENGGARA SURVIVE? LESSONS FROM THE POLICY IMPACT ON RESOURCE SUSTAINABILITY

Dede Rohadi

Sandalwood (*Santalum album*) plays an important role as a significant contributor to the income of East Nusa Tenggara District, Indonesia. During 1986 and 1987 to 1991 and 1992, sandalwood contributed around 2.5 billion rupiah (Rp) annually, or about 40 per cent of the gross product for the region.[*] Subsequently, the contribution declined to about 17 per cent during 1997–1998.

Oil is extracted from the heartwood and then exported to the US, Singapore and Europe for perfume and cosmetics. The wood is also used locally or on neighbouring islands (such as Bali) for woodcarvings and handicrafts in various forms (eg fans, pens, beads, rosaries, handbags). It is also used to make joss sticks for ritual purposes and is marketed domestically or exported, mainly to Taiwan. Currently, there are 4 sandalwood oil factories, 24 handicraft factories and 14 joss stick companies, with a total wood intake capacity of approximately 4000 tonnes.

The economic importance of the species has triggered the regional government to regulate each activity related to sandalwood production and sale. The main points of the current regulations are as follows:

- All naturally regenerated sandalwood (trees, dead trees and wood) belongs to the regional government. Any parties may plant the trees on their own land, but their share of the revenue from the harvested wood is set at 40 per cent (an increase from the 15 per cent that applied until 1996). Landowners must show their land certificate in order to claim their revenue from the sale of cultivated sandalwood.
- The regional forestry office conducts resource inventories every five years and determines the annual allowable cut for following years.
- The local government regulates harvesting activities, determines the harvesting cost allowance and specifies documents required during harvesting and wood transportation. In 1995 a harvesting cost allowance of Rp1300 per kilogram of heartwood was set, consisting of Rp800/kg for the cost of felling, bucking and skidding (which was the total wage delivered to the local community), and Rp500/kg for transportation, storage, monitoring, equipment, administration and levies.[**]
- The local government determines the wood price and allocates the wood to the selected companies. Following up on the 1995 example, class A sandalwood (the best quality) was priced at Rp18,000/kg, class B at Rp15,300/kg, class mix was Rp9000/kg, sapwood was Rp1000/kg and small branches were Rp500/kg. The wood was allocated to 20 companies.
- The regulations stipulate that all communities should care for the resource. Illegal cutting, stockpiling or transporting sandalwood, as well as intentional acts to damage the trees, will be prosecuted.

During 1997, the local government disallowed totally sandalwood cutting for five years (until 2002), due to the alarming rate of resource decline. One year earlier, the government launched an operation whereby all 'illegally cut' wood that was being stored by local communities was confiscated. This amounted to almost 2000 tonnes of wood, which was allocated to local industries to maintain their operations until the year 2003.

The regulations intended to sustain the resource have, in reality, had the opposite effect. The benefit of cultivating trees was insufficient, because only 40 per cent of the revenues, previously 15 per cent, were returned to households. The regulations promoted the active removal of young, less valuable trees growing in private fields in order to make space for agricultural production and to ensure that farmers did not have to deal with government functionaries, while valuable specimens were cut for illegal trade. The impact of the industry and regulations on resource stocks is illustrated in Figure 3.4. Official wood production (ie the amount reaching factories from the official allowable cut) fluctuated within the range of 87 tonnes to 995 tonnes per year between 1969–1970 and 1996–1997. Annual production during 1997–1998 was highest, at approximately 2000 tonnes, which was derived from the confiscation operation. Thereafter, this fell to zero due to the harvesting ban. Industrial capacity has increased significantly from only 800 tonnes per year in 1974. Capacity doubled by 1985 and exceeded 4000 tonnes by 1998. This increase was largely in the handicraft industries, which have been encouraged and promoted by government as traditional industries (eg carving and handicrafts). The big gap between industrial use and the official level of wood production is filled by illegal cutting. The estimated standing stock of sandalwood is based on inventory results and shows a sharp decline since 1994.

Currently, the resource base for the sandalwood industry is threatened. This illustrates a situation where policies to regulate the resource and promote industry have neglected community rights and where the impacts on individuals and communities have been adverse.

Source: Rohadi et al, 2000
* At the start of this period, US$1 = Rp1650, ending the period at Rp2200.
** At the time of study, US$1 = Rp2200.

costs and reduces the opportunities to achieve the economies of scale through larger-scale market operations and the linkage of markets over large geographic areas. Consequently, the benefits that arise from well-functioning markets (which, in turn, contribute to the integration of economic activities between regions), and the associated benefits to producers and consumers of access to larger-scale markets, are lost. Instead, there are fragmented and 'thin' markets that are not well integrated within the regional economy. In these situations, markets do not provide their desired broader economic benefits of contributing to specialization, economies of scale and increased income and employment by market participants. Instead, market participants may be tied to subsistence activities, which are known for their small-scale nature and a lack of specialization and exchange, while markets exhibit uncertain and unstable prices, contributing to risk and uncertainty for market participants.

Marketing and value creation: marketing as a productive activity

There can be a tendency for some people who manage primary resources, such as farmland or forests, to be particularly aware of values associated with those resources and to be dismissive or less aware of the values created in the marketing process. Typically, the complaints against 'middlemen' do not recognize the vital and productive role that they play in the process of exchange and distribution (Abbott, 1987). By making primary or forest products and services available when and where (and in the form in which) they are desired by consumers, the process of marketing creates value (or utility in economic terminology) by adding utility of time, form and place to the original forest product. Consequently, marketing is a productive activity by creating such values. As demonstrated in Box 3.1, the rewards may not be substantial. Nonetheless, they can be important sources of livelihood for people who have few alternative employment prospects. In situations where markets function well, the monetary values (unit prices) that are expressed at different stages of the marketing chain can be taken as a measure of these contributions and the costs that are associated with them.[3]

As implied in the preceding section, well-functioning markets offer participants the opportunity to adjust their patterns of household activities and production in order to achieve the economic benefits of specialization and exchange that are made available in the context of a larger market. However, if markets do not function well – for example, when there is opportunistic behaviour and consequent high transactions costs, or in cases of externalities or public goods – market prices may be unreliable measures of social benefits, as well as of production, marketing and consumption costs. This is in contrast to the private benefits and costs of these activities. This may occur with forest products that are derived from communal forests where customary practices to ration use may have broken down due to pressures on social or economic systems within the community, leading to unplanned depletion of the underlying resource. Another example arises from the sale of forest products in export markets if overvalued or undervalued exchange rates distort international price signals. A further example comes from the exertion

of market power in situations of monopoly or monopsony.[4] For instance, when analysing a market in order to assess market performance and to improve resource management or market operations, it is important to assess the nature and source of opportunistic behaviour, as well as the impacts of any externalities, (ie any extra-market benefits or costs) or other market imperfections that may be associated with the product and its extraction, use or sale.

Who performs the functions of marketing: market institutions

A fuller understanding of how markets work is gained if both a functional and an institutional focus are taken when studying them. The institutional approach attempts to determine the institutional components of marketing systems in order to answer the questions: 'Who are the people who are involved in the various stages of marketing, from the point of harvest to the point of consumption, and what are the institutional structures and associated operational rules?' This can be very effectively allied with a functional approach in order to understand the various marketing functions that are performed by the different institutions in the marketing system. The complexity of the relationships of different market agents and traders is illustrated in Box 3.3.

A typical classification of marketing institutions distinguishes the roles of marketing agents – those who perform some of the functions of marketing but do not take title to the goods they handle – from traders, producers and processors who take ownership of the goods in question. The distinction between these two categories of traders is important, since the incentives, risks and market functions for these two groups are generally quite different. Marketing agents include those who sell on behalf of others for a fee, salary or commission, as well as those who provide services such as transportation or facilitating activities to producers, processors and traders.

In contrast to marketing agents, traders take ownership of goods and resell to their best advantage, often accumulating products to the point of bulk sales (wholesaling) and subsequently dispersing them (here, traders may be known as brokers or jobbers). Consequently, traders contribute to the risk-bearing and exchange functions, in particular, and often to transportation and storage, among a number of other necessary contributions to marketing functions. Traders can potentially be classified by the scale of their operations (whether large or small). They may also be classified as travelling merchants, who sell at various locations, or as permanent traders, who sell at a single location.

Processors constitute another important category. This is also the case for facilitating institutions. Facilitating marketing institutions include those that operate price-discovery mechanisms or provide other services that facilitate the operation of a market. An example of a facilitating institution is CAMPFIRE (Communal Areas Management Programme for Indigenous Resources), which provides for the marketing of wildlife leases through rural district councils in Zimbabwe (see Box 3.8).

BOX 3.8 MARKETING WILDLIFE LEASES BY RURAL DISTRICT COUNCILS

Ivan Bond

CAMPFIRE (Communal Areas Management Programme for Indigenous Resources) aims to transfer responsibility for wildlife resources from central government to rural producer communities. The legal framework for CAMPFIRE was established when central government transferred the 'appropriate authority' for the management of wildlife to 12 district councils during 1989 and 1991. The incentives for both district and sub-district participation in CAMPFIRE are derived from the significant income that can be earned from wildlife resources. Since 1989, approximately US$10 million have been earned by the primary wildlife districts from leases established with private-sector safari operators. Over 90 per cent of this income is earned from the lease of sport or trophy hunting rights.

This case study examines the structure, conduct and performance of these lease agreements between the district councils and the private sector. The case is interesting because upon receipt of appropriate authority, district councils abandoned the controlled pricing methods of the wildlife department, preferring instead to allow market forces to play a significant role in the allocation of the wildlife leases.

Research methods

The analyses of the market for, and the allocation of, wildlife leases are based on a comparative analysis of the first 12 districts to receive appropriate authority. Central to the methodology was the active participation of the researcher in many process workshops across a wide number of districts. The qualitative nature of this component was supported by a quantitative analysis of primary and secondary economic data. To calculate the efficiency (performance) of the wildlife leases between district councils and commercial safari operators, an input–output index was developed (Child and Bond, 1994; Bond, 1999). The input was calculated by applying a set of standard values to the sport-hunting quota issued by the wildlife department. The value of the output was the gross revenue received by the district council from the use of that quota. The index allowed quantitative comparisons to be made between districts and between years within districts. In addition, an overall efficiency index was calculated and compared with the free market prices obtained on an annual basis from the Zambezi Valley Auction Hunts (ZVAH) (see Child, 1995). Average trophy prices, received by the wildlife department and district councils, were also compared.

Results
Market structure
The structure of the market for sport-hunting leases in the communal lands is determined by: the number of lease areas for sport-hunting leases in districts with appropriate authority (supply) and the number of commercial safari operators wishing to operate in those leases (demand). Over the period of 1989 to 1993, the number of district councils with appropriate authority increased from 2 to 12. However, in some districts there were more than one lease area, each with a discrete sport-hunting quota. Consequently, the number of leases in the communal lands increased from 4 in 1989 to 20 in 1993.

The changes in the number of registered tour and safari operators have been used as a proxy indicator for demand by the private sector for sport-hunting leases. In 1985, there were 60 registered tour and safari operators who were members of the Zimbabwe Tours and Safari Operators Association. By 1991, this number had increased to 209. However, these operators were not restricted to hunting in the districts with appropriate authority. They could also compete for leases in protected areas managed by the wildlife department and for plains game leases on commercial ranches. Some of the registered operators would also have been landowners. It has been shown that access to 'big game' significantly increased the financial viability of operators (Child, 1988). It can therefore be assumed that most of these operators wanted access to sport-hunting wildlife leases in the communal lands or state protected areas where they would have access to key species.

Market conduct
For the marketing of sport-hunting leases to commercial safari operators, it is assumed that the conduct of the seller (the district council with appropriate authority) was the primary factor determining the conduct of the market.

If, in the long term, CAMPFIRE aims to achieve full and functional institutional change for the management of wildlife at the producer community level, then the producers must be involved in the decisions pertaining to allocating leases. An examination of ten case studies showed that the involvement of wildlife producer communities in the marketing of leases varied considerably between districts (Bond, 1999). Generally, the district executive officers controlled the process due to their education, positions of responsibility and power. The cases, however, where there were producer community involvement generally resulted in efficient and accountable contracts.

Performance

The analysis of the annual average efficiency of wildlife resource use by district councils showed that between 1989 and 1993 efficiencies increased by 11 per cent (see Table 3.3). Between 1990 and 1993, however, the index increased from 0.35 to 0.50, which could be considered more representative due to the low number (n = 2) of districts that had appropriate authority in 1989. The annual average is approximately 30 per cent of the annual index of efficiency of wildlife resource use derived from the Zambezi Valley Auction hunts. This significant difference is due to the quality of wildlife habitats. In general, wildlife habitats in the communal lands are fragmented while the Zambezi Valley is characterized by a continuous block of protected land in excess of 10,000 km^2.

Table 3.3 *Comparison of average annual efficiency of resource use within CAMPFIRE districts and the free-market prices of the Zambezi Valley Auction Hunts (ZVAH) (see methods)*

	1989	1990	1991	1992	1993	% change
Annual average efficiency of all districts	0.45	0.35	0.34	0.40	0.50	11%
Annual average efficiency of ZVAH	1.36	1.35	1.36	1.46	1.59	17%

The input–output efficiency index is unable to differentiate between the changes in efficiency as a result of price increases, the actual success rate of hunting, or both. The analysis of elephant and buffalo trophy prices received by district councils and the wildlife department has been used to examine the impact of price changes. Although average annual trophy prices were higher in the state-protected area (ZVAH), communal land trophy prices increased significantly over time (see Table 3.4).

Table 3.4 *Comparison of average annual trophy prices paid to either the district councils or the wildlife department for trophy elephants (US$)*

Area/species	1989	1990	1991	1992	1993	% change
Elephant						
Wildlife department (WD)	10619	9546	8716	8577	12028	13%
District councils (DC)	2959	6275	5013	6974	7470	152%
Ratio WD:DC	3.59	1.52	1.74	1.23	1.61	
Buffalo						
Wildlife department (WD)	2309	1904	2539	4970	1992	–14%
District councils (DC)	443	408	293	459	926	109%
Ratio WD:DC	5.21	4.67	8.67	10.82	2.15	

Discussion

The results show that devolved management of wildlife has allowed district councils to benefit from new methods of marketing wildlife resources. The extent to which wildlife producer communities have been involved, and the strength of their proprietorship over the process, are at this point uncertain.

Some features of rural markets

Much information that is available on rural markets has resulted from the studies of economic geographers and anthropologists. These studies have been supplemented by the studies of ethnobotanists and have provided valuable delineation of spatial, functional, cultural and social characteristics of many local markets (see, for example, Cunningham, 2000; Martin, 1995). Some of these studies have identified spatially based hierarchies of markets at different levels of functionality within the marketing system. This occurs when locally sourced resource goods move, for instance, from rural resource areas through local markets, to intermediate-sized markets and subsequently to central markets located at, or near, large consuming centres. An opposite movement of imported or processed goods from cities to rural areas has also been observed and described by economic geographers.

Traditional or periodic markets are often found in rural areas. These markets operate at an appointed location at which people gather to buy, sell, or interact socially (Bromley, 1980). Periodic markets occur regularly (eg weekly) at long-standing locations and may be a major means for small-scale farmers and artisans to find markets for their products. They also may be a major source of access by rural households to locally produced goods. In slightly larger centres in rural areas there will be traditional permanent markets, held daily, which also perform these roles; here, permanent sellers maintain stalls (Figure 3.5). Processed domestic foods and household products may also be available for purchase in a local store in some rural locations, but the range of store goods may be sparse.

In traditional markets, rural people usually bring goods that they have harvested, produced, or constructed for exchange with or sale to other rural people and other buyers. Traders may also bring goods that they have purchased for resale at local markets. Traditional markets may not have major facilities but customarily occur in long-established locations. Exchange occurs on the basis of private treaty involving inspection, typically with immediate sale and settlement. Some specialized rural markets, directed more at tourists and travellers than at other rural residents, are evident on roadsides or near tourist attractions (see Box 3.4).

During the 1960s and 1970s, numerous studies were conducted in traditional markets that challenged some widely held views that the activities of rural traders were exploitative and unproductive. Studies of the temporal and spatial relationships of prices, mainly for food, were conducted in order to assess these dimensions of pricing efficiency.[5] The results were mixed. Comparisons were complicated by a lack of pricing transparency and because differences in prices could be hard to assess, due to a lack of standard measures and grades. Even so, numerous studies that compared prices in spatially separated local markets concluded that there was a tendency for traditional markets to exhibit price variations, reflecting transportation cost differences between spatially separated markets. When temporal price patterns were assessed, there were also price variations that more or less reflected storage costs within a year . Such findings have been reported for traditional markets in both Asia and Africa (Riley and Weber, 1983). Most of these studies assessed the

Photograph by Manuel Ruiz Pérez

Figure 3.5 *A permanent market in Cameroon – Mokolo market. The market deals in numerous forest products, including a range of wild foods and medicinal plants*

relationships over space and time for market prices of non-perishable, storable foods (such as rice or other grains).[6]

Despite indications that the 'law of one price' often applies within regional markets for non-perishable goods, there are concerns that traditional markets may not always be efficient (Schaffer et al, 1987). Efficient markets require transparent prices. Private treaty trading, whether conducted in rural markets or in other contexts, does not lend itself to accurate price reporting. A lack of defined measures and grade standards does not add to price transparency. If markets are thinly traded,

with small numbers of sellers or small volumes, and if few buyers are attracted to these markets, prices will be volatile and hard to predict, contributing to uncertainty in planning by market participants. In these circumstances, poor communication and the lack of transportation may discourage suppliers and buyers and limit the scope and size of the local market.

Unpredictable, unreliable markets do not encourage investment in market infrastructure. These markets do not encourage the development of specialized market agents (such as brokers, agents, and facilitators) who may contribute to larger markets that have more scope for specialized activities. Consequently, the growth in exchange and the growth in market and employment opportunities that can be available in larger regionally integrated markets are not achieved in most local markets (Schaffer et al, 1987).

One of the major problems limiting the expansion of markets and trade in non-industrial economies is the lack of infrastructure to facilitate transport and communications (Ahmed and Donovan, 1993). High marketing costs reduce the possibility of local and regional trade and limit the economic linkages between urban centres and rural communities. The emphasis on reduction in government expenditures that has accompanied structural adjustment policies appears to have contributed to the lack of investment in communication and transportation infrastructure in the 1990s. Governments may try to stimulate local value-add by simple means, such as by export bans on unprocessed products (see Box 3.5). However, successful development of local processing requires adequate infrastructure in the form of transportation and communication, in addition to management capability, access to capital and knowledge of buyers' preferences and how to cater to these.

Any seasonal variations in supply or demand are likely to be reflected in seasonal price variations for the products in question. Seasonal price differences for such products should not be confused with a price trend which could reflect more fundamental changes in supply or demand, such as an increase in demand arising from a higher consumer income or from scarcity associated with depletion of the resource. An example of the impact of seasonality on product availability has been noted by Cunningham (1990; 2000) for palm wine in an area of south-eastern Africa. When the popular local wild marula fruits (*Sclerocarya birrea*) were available, commercial sales of palm wine fell as households brewed their own preferred substitute. However, at Christmas time, when migrant workers were home, palm wine sales increased.

Analysing markets: marketing efficiency and performance measures

The concept of efficiency

How do we evaluate how well a market works? Well-functioning markets help to maximize welfare from marketing processes and help to minimize marketing costs. The long-recognized economic approach to market assessment relies on the concept of marketing efficiency. Conceptually, marketing efficiency comprises two fundamental components: pricing and technical efficiency (see Box 3.9). Technical efficiency relates to the least-cost way of organizing the functions of production and marketing in order to achieve the most cost-effective means of satisfying consumers' preferences and demands. Examples of technical efficiency are the use of appropriate and cost-effective technology and procedures in transportation, handling and storage in order to reduce spoilage and wastage from the point of harvest to the point of consumption. In contrast, the concept of pricing efficiency relates to the pricing characteristics associated with an efficient and well-functioning marketing system. The criteria that could be used to judge whether or not a market displays pricing efficiency include the following:

- Are consumers' preferences reflected in the prices that are transmitted through the marketing system, so that the variety and types of products and services that are preferred by consumers are the types of products and services that are, in fact, harvested, produced, processed and marketed?
- Do the prices of goods and services reflect actual or potential competition in the provision of marketing services and functions, so that costs associated with marketing are not inflated by inefficient practices or excess profits (rents)?

Although there may be many circumstances in which pricing efficiency and technical efficiency are complementary concepts, there are instances when these two influences may conflict. Examples arise when the attempt, by individual firms, to achieve economy of size is limited by the minimal demand of regional or national markets for a manufactured good. In circumstances such as this, the available economies of scale that are involved in manufacturing could be exhausted by only one or two firms. However, the existence of one or two firms

---

BOX 3.9 CONCEPTS TO ASSESS MARKETING EFFICIENCY

Market(ing) efficiency reflects a combination of:

1   technical efficiency and the growth over time of the marketing system and its components (ie productivity); and
2   pricing efficiency, reflecting:
   - whether prices and the markets that generate these are responsive to consumers' preferences; and
   - whether the prices of goods and services reflect the minimum levels of costs (implying competition in providing marketing services that drives down monopoly rents and reduces inefficiencies in marketing processes, as well as the absence of external technical economies or diseconomies, ie externalities; these may arise if inputs are unpriced or if their prices do not reflect social costs).

---

could compromise pricing efficiency, even though an industrial organization policy that maintained only a few firms could be technically efficient.

Industrial organization concepts relating to market performance

Economists have devised a number of formalized theoretical models of economic behaviour that relate to characteristics of market structure and market conduct, with reasonably predictable market outcomes. These models (described as perfect competition, monopolistic competition, oligopoly models of various types and monopoly[7]) are stylized representations of economic behaviour. The economic theory of the impact of firm behaviour on industry outcomes, known as industrial organization theory, provides some insights that may be useful in assessing how markets work and how market performance may be assessed. This section addresses these issues.

The classical paradigm of industrial organization analysis relates market structure and the behaviour of firms that operate in a particular market to market performance, as depicted in Figure 3.6. Following Bain (1969) and others, the parameters of market structure are consid-ered to involve the number and size distribution of competitors; the barriers that may limit entry and exit; whether the product is homogenous or differentiated; and whether there are substitutes for the product. Other structural factors relate to the existence and extent of vertical integration; the availability of information; the nature and extent of risk that confronts economic agents in the particular market; and characteristics of demand, such as the existence of consumption substitutes.

The conduct dimensions of this paradigm relate to the behaviour of economic agents in the industry. Specifically, market conduct focuses on whether firms act independently or interdependently in their output, pricing and product (ie non-price) decisions, and the nature of these decisions. The nature of the interdependence between firms that applies in actual markets is of importance. This may involve different forms of rivalry or different forms of cooperation. Some of these types of behaviour may, in turn, influence structure, while market structure and conduct influence industry performance. Market performance, reflecting resource allocation at the industry level, is often assessed in terms of measures of profitability; whether economies of scale are exhausted and other efficiencies are

93

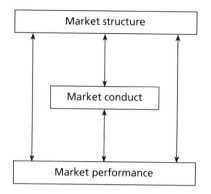

Figure 3.6 *The industrial organization paradigm*

achieved; whether innovation is stimulated; and the nature of the variety and product choice that confront consumers (Bain, 1969; Green, 1990; Jacquemin, 1987; Scherer and Ross, 1990). These criteria are noted in Table 3.5.

Recognition that the concept of perfect or 'pure' competition is a useful conceptual abstraction, rather than a description of real world markets, led to the concept of 'workable competition'. This applies when there is rivalry among sellers, sellers try to maximize profits and price discretion is limited by the option of buyers to purchase from rival sellers (Clark, 1940). Thus, when there are several sellers of a particular product, driven by profit goals, and when easy entry conditions for new sellers keep established firms honest, workable competition ensues (Green, 1990). The importance of 'entry conditions' to firm conduct and industry performance is also recognized by a more recent concept of industrial organization theory: contestable markets. Baumol et al (1982) have argued that the welfare properties of perfect competition can be achieved under conditions of oligopolistic markets as long as there are no impediments to firm entry and exit.

In practice, market competitiveness can be viewed as a matter of degree; the conditions of entry of potential competitors to any market or industry are viewed, by economists, as influencing market conduct and performance. Cost advantages may accrue to existing firms because of economies of scale, learning by doing or customer loyalty. Incumbents may also have an advantage because of experience with local culture, customs or language. Natural barriers to entry, due to geography, may be overcome (partially) through transportation and communication; this limits the extent to which existing firms can raise prices above their average cost. Governments may also create barriers to entry. Examples include quotas, permits, preferential purchasing and patents. Competitiveness is aided if industry and markets remain open to new entrants, rather than restricting or regulating entry. Improvements in transportation and communication infrastructure will increase market competition. Low levels of economic protection to local industry, through low levels of tariffs and the removal of non-tariff barriers, can act as a surrogate for easy entry, providing competition from imports for firms that operate in fairly concentrated markets.

Documenting marketing margins, channels and returns within the marketing system

In view of the lack of knowledge and the lack of published data on the operations of local markets, field research methods that include participatory involvement of researchers with key informants who represent, or are knowledgeable about, the industry (such as suppliers, producers, traders and other participants in the marketing chain) are necessary in order to identify major marketing channels, issues and problems for many local markets. Chapter 6 describes participatory approaches in further detail. As pointed

Table 3.5 *Market structure, conduct and performance: major measures*

	Characteristics
A Market structure	Number and size distribution of firms (concentration ratios)
	Barriers to entry: capital costs; economies of scale; conduct of firms; regulations
	Nature of product: homogenous or differentiated; existence of substitutes
B Market conduct	Independent or interdependent actions
	Coercive or collusive behaviour
C Market performance	Profit levels: are these excessive or 'normal'?
	Cost levels: are economies of scale exhausted; is there inefficiency in operations that reflects a lack of competitive pressure – sometimes termed 'x-inefficiency'?
	Innovation within the firm or industry
	Product choice available to consumers

out by Trager (1995): the basic (ie minimum) questions to be considered in studying exchange and distribution systems include the following:

- What is being exchanged or distributed?
- Who is engaged in these exchanges?
- How are exchanges organized?
- When does exchange and distribution take place?
- Where does exchange and distribution take place?

A useful collection of case examples where various market research tools are applied to particular market issues and problems is given by Scott (1995). Examples of particular studies of markets that use participatory or rapid appraisal methods are reported by Morris (1995), Holtzman et al (1995) and De Jong et al (1999) (see Box 3.5). A summary of research procedures used to explore the role of indigenous rural markets in two areas of Nigeria is given by Porter (1995), who outlines the need for pre-field library, cartographic and archival research, as well as acquiring necessary regional and local permissions. Initial field-level activities include identifying the location of

markets, from printed schedules and key informants, and constructing and testing interview guides/surveys. This requires pre-testing and subsequent application, with the help of field assistants who must be recruited and trained. A more general overview of economic concepts and the tools of market analysis is given by Scarborough and Kydd (1992). An overview of market research tools, which takes a business-oriented focus, is given by Crawford (1997). Table 3.6 illustrates some of the measurements of product attributes and other data recorded in a study of woodcarving markets in Zimbabwe (see Box 3.4).

The appropriate analytic approach and method to be used in any market analysis should, of course, be considered in the context of the particular market issue or problem, and this should be identified before survey design and application. A call to use standardized methods and measures in collecting data and reporting studies is given by Moran (1995). The importance of analysing a standard set of marketing-channel descriptors for forest products is presented in Box 3.10.

A first step in describing markets, applying methods noted above, is to

Table 3.6 *Data collected through a market monitoring system for woodcarvings in Zimbabwe (see Figure 1.5). Three markets were monitored, and all carvings were tagged. The variables recorded for each species are shown*

Variable	Units	Method
Length, breadth, height	Cm	Callipers used to measure carvings; data recorded by market enumerator
Expected price	Z$	The price the retailer expects from the sale of the carving
Floor price	Z$	The lowest price for which the retailer would sell the carving
Actual sale price	Z$	The actual sale price of each carving
Date into market; date out of market (sold)	Day/month/ year	The date each carving entered the market and the date each was sold
Description of carving (product type)	For instance, hippo, giraffe and warthog	Recorded for each product by the market enumerator
Species	*Shona* name of species for each product	Cross-referenced to scientific and common names

Note: The study is summarized in Box 3.4.
Source: Braedt and Standa-Gunda, 2000

follow the basic questions posed by Trager (1995). These questions identify the major marketing channels for a product from the stage of initial production, or harvest, until it reaches the final consumer. It will be useful to identify who performs the various functions, the manner in which they are performed and, where possible, the margins and returns that accrue to participants. The people, organizations and institutional arrangements involved in performing the various marketing functions within the marketing system can be identified from interviews with key informants – for example, from participatory surveys of a representative sample of market participants. This will highlight the basic characteristics of the marketing system. Procedures used to document marketing channels and margins for commercialized crops in a region of Mexico are discussed by Mendoza (1995).

Calculation of a simple gross marketing margin (consumers' price less producers' price) is typically reported as a percentage of consumers' price.

Alternatively, levels of unit marketing costs, and the components of this, can be surveyed, allowing the net margins (net of marketing costs) to be calculated and reported, as with value-added estimates (final selling price of the finished product less cost of all purchased inputs; see Box 3.4). These measures require detailed information from market participants. This, in turn, raises questions about the accuracy and reliability of the price and cost data that may be elicited, highlighting issues of sample size and sampling procedures. Price surveys that are conducted at various levels of the marketing channel can provide information on the stability or variability of prices and the distribution of the gross, or net, marketing margin at different levels of the marketing channel. Some price data may be available from government sources. The quality of available price data and its applicability to the marketing issue that is of interest to the researcher should be assessed. For example, in an assessment of the implications of maize market liberalization in

Box 3.10 Comparative analysis of forest products: a proposed methodology

Brian Belcher and Manuel Ruiz Pérez

A rich body of information and experience on various aspects of the development of NTFPs has emerged over the past two decades (eg de Beer and McDermott, 1996; Cunningham, 2000). Numerous case-based studies of elements of forest product systems have been prepared. Many interventions directed at development have been tried at the project level, including combinations of technical, institutional and financial support for forest-product production, processing and marketing. Most have met with mixed success. Larger cross-cutting interventions have been attempted, including 'green markets', 'fair trade' initiatives and efforts to promote NTFP certification.

Even so, the major elements that may underlie project success or failure are unclear. The assessment information has been gathered using a variety of methods (often poorly described), at different scales, focused on different elements of the forest-product production, processing and marketing systems. Work is needed to document and compare cases using consistent terms and definitions for an appropriate range of variables.

The Center for International Forestry Research (CIFOR) is developing a method to synthesize lessons from approximately 45 cases (15 each in Latin America, Africa and Asia), that have already been researched and analysed, by applying a uniform comparative analytical approach. The comparative method is based on that developed by Ruiz Pérez and Byron (1999). It uses multivariate analysis techniques to find patterns, to develop typologies, to identify key context variables and to analyse their relationships with observed development outcomes.

A range of descriptors (variables) has been identified. The selected cases are described according to the standard set of descriptors to develop a case-study matrix. In order to capture the relevant variability, the analysis will consider the whole system, from production of raw material through to final market, including social, economic, technological and ecological aspects of the production systems, of the products and of the market. The following categories of information will be addressed:

- geographic setting;
- biological and physical characteristics of the product;
- characteristics of the raw-material production system;
- socio-economic characteristics of the raw-material production system;
- institutional characteristics of the raw-material production system;
- characteristics of the processing industry;
- characteristics of the market and marketing system;
- outside interventions (policies, subsidies, projects, etc);
- outcomes of forest-product commercialization.

As an example of the depth of description, the variables for one of the above categories, 'characteristics of the market and marketing system', are as follows:

- age of market – how long has the product been traded;
- market trend – has the market expanded, remained stable or contracted in the past ten years;

- total number of commercial raw-material producers;
- total number of traders in the case area (of various levels);
- total number of processors;
- size of the trade – what is the value at various levels;
- market transparency – degree to which raw-material producers know about the marketing chain;
- importance of 'vertical integration' – degree to which processing firms have owner-ship in firms supplying their raw material or in export and marketing firms;
- level of organization among traders – is there a formal trade organization and what is its age;
- degree of participation in the organization;
- barriers to entry – are there barriers that make it difficult for new traders to enter the industry;
- intensity of state involvement affecting forest product trade (regulations, incentives).

Exploratory data analysis techniques are then proposed to outline patterns, gradients of variability, clusters of cases and key variables associated with them (see Figure 3.7). Essentially, the aim is to develop a useful typology of cases, identify key variables (those with maximum explanatory power) and investigate relationships between particular classes of forest-product production. It is important to identify which sets of character-istics, or 'types' of cases, tend to be associated with what kinds of human development and conservation outcomes. This information should make a valuable addition to the management and policy debate.

Kenya, Ngugi et al (1997) were able to apply cointegration methods of price analysis to official price series for maize, collected for the construction of the national consumer price index.

Broadening the scope of sampling and of recording spatial and temporal prices can provide information on regional price variations, as well as information on systematic and erratic (or random) differ-ences in temporal prices. These differences may reflect some or all of the following:

- patterns of seasonality in prices;
- price cycles that may be associated with changes in production patterns over a number of years; and
- long-term trends in prices.

This type of price analysis, as applied to rice and maize in Ecuador, is discussed by Tschirley (1995). Identification of market-ing channel participants and marketing margins will not provide all of the infor-mation needed to assess market performance. It will, however, provide basic information on some features of the marketing system. Since cultural affinities can be important components of market networks (see Box 3.3), such economic issues should be assessed in the light of their social and cultural context. As a result, a multidisciplinary approach that identifies the participants and features of local and regional markets is likely to be most informative in obtaining a fuller understanding of the marketing system for a particular product or group of products. Chapter 7 further considers the integration of economic perspectives with those from other disciplines. A multidisciplinary approach may also help to identify opera-tional rules, whether formal or informal, that govern the behaviour of those who participate in the market.

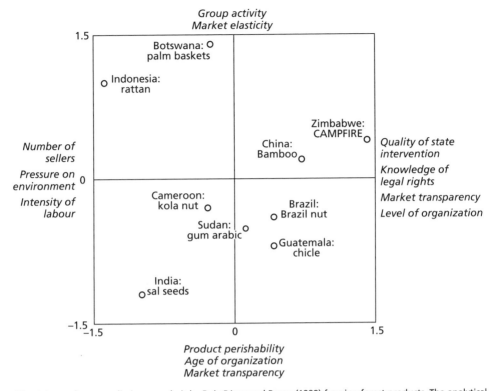

Figure 3.7 *An ordination diagram showing the relationship among forest products described by a standard set of descriptors*

Note: The data are from a preliminary analysis by Ruiz Pérez and Byron (1999) for nine forest products. The analytical method (such as non-linear principal components analysis – PCA) displays the products so that their similarities and differences can be evaluated. Those products placed closely on the diagram are similar to each other, whereas those widely separated are distinctly different. The descriptors that are associated with particular axes are also shown. In this case, for instance, the Indonesian rattan case is indicated to be completely different from the Zimbabwean CAMPFIRE case. The former is characterized by many sellers, high pressure on the environment and high intensity of labour (and the CAMPFIRE case has the opposite characteristics).
Source: adapted from Ruiz Pérez and Byron, 1999

Analysing market prices

More complex assessments of price relationships between regions and over time can be performed to assess regional and temporal aspects of market integration and price linkages. These require relatively large amounts of reliable price data. Supplemented by reliable data on quantities that are produced and consumed, together with other information on the structure of particular markets, economists have developed and applied models of trade flows, demand estimation and the assessment of purchasers' preferences.

As was noted in the section on 'Some features of rural markets', time-series or cross-section (spatial) price observations have been analysed in order to assess aspects of pricing efficiency related to 'the law of one price'. In their simplest form, such tests assess whether prices in different markets are correlated statistically, or whether the price differences (margins or spreads) that are exhibited between spatially separated markets are relatively stable. These relationships can suggest whether such markets are highly integrated, as one would expect if arbitrage occurs. In the same vein, if price differences between particular markets reflect the costs of transferring products between those markets, it could also be inferred that markets are relatively competitive. Furthermore, if the levels of costs that separate markets spatially are as low as possible, technical efficiency could be inferred. However, evidence from simple tests of price-series correlation may not always reflect the existence of market competition or market integration; instead, results may reflect the influence of inflation in the general level of prices, or trends in population or income that are reflected in fluctuating demand (Badiane, 1999). With reliable price series, economists have sought to test for patterns of associations in prices, employing more sophisticated 'time-series' analysis that focuses on various structures of 'autoregressive' lags in different price time series. This type of study tests for patterns of statistical association that may imply patterns of causality and seeks to identify the nature, or 'direction', of such causality. An early example is given by Ravallion (1986) for rice in Bangladesh. Studies that test for 'cointegration' of prices may also be allied with structural analysis of markets. An example of time-series analysis of prices for woodcarvings, reported in three nearby craft markets in Zimbabwe, is outlined in Box 3.4.

Other approaches to analysing markets

A variety of other analytical approaches may be also applied that focus on marketing opportunities for particular groups of producers or for particular products. These could include, for example, assessments of potential demand and business plan analy-sis. Reference to standard texts on marketing methods, marketing analyses and business planning will indicate alternative methods that may be applied to assess particular marketing issues and problems.

Another approach to analysing markets is to focus on the implications of policy interventions. Studies that take this focus have been used in the agricultural sector. One tool to analyse agricultural commodity markets along these lines involves the concept and methodology of the policy-analysis matrix approach developed by Monke and Pearson. This approach is described and applied by Shapiro and Staal (1995) and Staal (1996) for dairy in regions of Kenya and in Addis Ababa, respectively. The procedure focuses on profits, the difference between revenues and costs, and attempts to assess divergence between private (ie market) and social (ie socially optimal) values. Estimates are arrived at for each input and output relevant to the activity that is the focus of the study (eg the production and sale of milk). A policy-analysis matrix will develop estimates of the absolute levels and divergence between private values and social values for revenues and costs (for tradeable and non-tradeable goods), and will relate these to estimated costs of subsi-dies, taxes, duties, rent-seeking and market failures. Accurate detailed budget data are needed for the various activities in this macro-level approach.

The Organisation for Economic Co-operation and Development (OECD) has adopted a somewhat related approach to summarize features of policy impacts on the marketing system of the industry sector, more commonly used to assess the aggre-gate impacts of national policy on farm sectors in high-income nations. This involves calculating (based on agreed procedures, including the basis for specify-ing comparison of prices) the transfers to agricultural producers that are achieved through various government-based policies, and also the implicit consumer-level taxes that may be the means of achieving these transfers. The resulting estimates of producer and consumer subsidy equivalents are used to monitor and compare national policy interventions.

Policy-oriented analyses may also be based on the collection and analysis of disaggregated data at the level of the household or for individuals. An extensive body of literature has been developed, for example, on methods for household surveys related to nutrition and food availability and for labour use studies. Examples are outlined by Von Braun and Puetz (1993).

Summary and conclusions

This chapter has provided an overview of some basic economic concepts that relate to the markets for forest products. Basic approaches to analyse markets and to assess market performance were outlined. Detailed knowledge of the structure, conduct and performance of local and regional markets for most forest products is lacking. The documentation of marketing channels, institutions and functions, allied with analyses of marketing costs and returns, will fill in gaps in knowledge about the market systems for forest products. Knowledge provided by market participants of the problems that they face will help to explain how these markets work and the nature of the important potential contributions of these markets to the livelihoods of rural people. Understanding and documenting market performance for forest products may aid in developing policy or interventions that may, in turn, improve the structure, conduct and performance of markets. More knowledge may contribute to appropriate resource management policy that is oriented towards assisting individuals, households and communities in improving and sustaining their access to employment and income derived from forests.

Notes

1 Marketing chains, which link primary producers or harvester-gatherers with final consumers, are sometimes called marketing channels; the multiple groupings of the different marketing channels that may link groups of producers and consumers are often called marketing systems.

2 Since the benefits of public goods are not reduced by any one person's consumption of these, and it may be neither possible (nor desirable) to prevent more people from consumption, there is little private incentive to provide such goods. Consequently, public goods typically require collective provision, as through government.

3 These monetary values may also reflect profits or rents that relate to resource scarcities or the returns to entrepreneurship or risk-bearing.

4 That is, where the ability to control market supply or demand provides single sellers (monopolists) or single buyers (monopsonists) with effective power to influence market prices. Appropriate public policy in these instances may encompass competition (antitrust)

legislation and/or policy or other ways to improve market competition – for example, through reduction of non-tariff or tariff barriers to encourage import competition.

5   The appropriateness of these features as a complete measure of pricing efficiency is debatable. Pricing patterns that reflect transportation and storage costs may be expected in perfectly competitive markets. However short-term variations in supply and demand may obscure this (see the 'Summary and conclusions').

6   This type of price pattern is not expected if: market information about prices is not readily available to buyers and sellers, when there is little information about the quality of goods for sale, and when markets are small and there are few traders. These circumstances apply for sales of the medicinal bark of *Warburgia salutaris* in Zimbabwe. Prices for unit quantities of this product – based on purchases from and interviews with, sellers of medicinal bark in markets in major centres and intermediate markets in Zimbabwe during May 1999 – did not demonstrate a spatially consistent pattern of prices (Veeman, M et al, 2001)

7   Perfect competition refers to an industry in which there are large numbers of firms (ie sellers), each of which acts independently, with full information, in producing and selling a homogenous product. No firm can individually affect price levels. In the long run, increases (decreases) in market prices in excess of (below) average costs will attract new entrants (cause some firms to leave). This causes supply to adjust to the point where costs are just covered. In monopolistic competition, however, each of the large number of small firms may have some ability to influence selling prices, and asking prices may vary somewhat since each firm sells a slightly different product or service. In oligopolies (oligopsonies), the sellers (buyers) are fewer in number and larger in size; firm and industry conduct and performance are more difficult to predict. In an industry with a monopoly (monopsony), there is, by definition, one seller (buyer) of a product that has few if any substitutes. Depending upon the nature of the market demand (supply) that this firm faces, there may be significant departures from pricing efficiency and the potential for the firm to earn excess profits.

An introduction to approaches and issues for measuring non-market values in developing economies

Peter C Boxall and Tom Beckley

Introduction

In today's globalizing economy, it is important to attempt to understand and document the full value of forest resources. Transformation of forest resources into marketable products (with prices) may not be the highest and best use of trees and other forest resources. This is particularly true in the developing world where these resources directly support rural livelihoods or form an integral component of unique cultures. Without the ability to compare market and non-market values, however, we are not able to determine whether resources are allocated to their highest valued use in specific local contexts.

Recently, there has been great interest in eliciting full values of market and non-market, priced and non-priced, goods and services related to forests and trees in the developing world. This chapter is intended to familiarize readers with some of the methods available for estimating non-market values. There is a vast array of non-market valuation tools available. Each has its own advantages and disadvantages. While space available here does not allow for a comprehensive treatment

of all methods, nor the complexities involved with implementing them, the reader should gain an appreciation for the current state of practices.

The task of valuing non-market goods and services in developing world contexts is vast and the challenges are many. This is especially true for valuing things such as biodiversity (and the potential of species-rich forests to provide future scientific and medicinal breakthroughs), sacred and cultural values for trees and forests, and the value of ecosystem functions, such as air and water purification. Non-market valuation methods have a great potential to provide valuable information at the household level. A number of non-market valuation methods are appropriate and useful for eliciting the many ways that forest and trees contribute economically to household livelihoods.

Many of the tools and methods for quantifying non-marketed goods and services have been formulated in developed countries and transferred to the developing world. This leads to additional challenges in bringing the non-market valuation tool kit to developing world contexts. The developed world has functioning markets and many more marketed goods and services which make it easier for respondents (conceptually) and researchers (operationally) to convert non-market values into a common currency. In developing nations, however, markets for goods and services may be 'thin.' The reasons for this are twofold. Firstly, property rights for the resource in question may not be well defined, or may be defined in such a way that the resource is available for all but cannot be sold. This situation describes a wide range of goods and services that are involved in subsistence and semi-subsistence activities. Secondly, individual households may not be able to produce more than they can use, with the result that there is no surplus production to sell in a marketplace. In

some cases, surpluses exist, but such surpluses are bartered rather than sold in exchange for a common currency. These features add challenges to valuing non-market goods in the developing world.

It is arguably more important to perfect non-market valuation methods in the developing rather than the developed world. The dependence on natural resources, in general, and forests and trees, in particular, for rural livelihoods in the developing world means that a greater range of goods and services there are 'unpriced'. This lack of pricing information results in less data and information available to policy-makers for making critical resource allocation decisions. Providing this information would be useful in setting development priorities, evaluating individual projects (see later in

Photograph by Francis Ng

Figure 4.1 *A logging camp in Malaysia – with the loss of the forest there are a number of forest benefits that dwindle or disappear. Changes such as loss of biodiversity, loss of soil protective benefits and altered river flow are exceptionally hard to value. Some of the methods outlined in this chapter may be used for the valuation of such forest goods and services*

this chapter) and understanding more fully the linkages between human preferences and the environment (see Figure 4.1).

Perhaps one of the most compelling reasons to measure and assess non-market values is for cost-benefit analysis (CBA; see Chapter 5). In this technique the benefits of a project or programme are enumerated and compared to the costs in order to assess the economic efficiency. A project 'passes' the CBA test if the benefits are greater than the costs. An issue with this calculation, however, is the assessment of benefits and costs that do not show up in the economic market – in other words, the issue involves the identification and measurement of non-market values. Frequently, these values are ignored in the economic calculus, leading to erroneous CBAs and sometimes the completion of projects that are unjustified on economic grounds.

This issue is critical in developing countries because of tensions between environmental values and development values (Georgiou et al, 1997). The debate between environment and development frequently concerns the relatively high value put on development projects to solve problems such as malnourishment, unemployment or underemployment. However, the services provided by forests to local peoples, for example, or the attachment of the local cultures to forests, can be absent in this debate. Georgiou et al point out the tensions in these discussions and identify non-market valuation as one solution to incorporating these concerns in development policy contexts:

'Bringing discussion of rights and intrinsic values into the policy dialogue can be counterproductive in such contexts: honouring them is perceived as forgoing the benefits of development. If, on the other hand, conservation and the sustainable use of resources can be shown to be of economic value, then the dialogue of developer and conservationist may be viewed differently – not as one of necessary opposites, but of potential complements.'

Some view environmental concerns as luxuries that need to be deferred until a later stage of development. This view is especially germane when the non-market values of forests in a developing nation accrue largely to individuals residing in the developed world. From a policy perspective, the critical issue is not the valuation of a given resource per se, but rather who benefits and who pays when resources are used for different purposes. Developing nations are not likely to institute policies that protect environmental amenities, or for values associated with the mere existence of these amenities, unless they see some immediate payoff, given the challenges in daily survival and low relative standards of living. In these contexts the challenge is for developing nations to 'capture' the conservation benefits in the form of cash income or technology to assist in development efforts. In these cases complementarities between conservation and development initiatives are clearly of value. Thus, economists interested in non-market valuation must reveal and evaluate these values (the demonstration process) and, secondly, find ways for the developing country to capture these values (the appropriation process) to enhance the well-being of its citizens.

Having set the stage for why non-market valuation may be important in developing-world forest management contexts, this chapter proceeds in the following manner. Box 4.1 provides readers with some background economic

BOX 4.1 ECONOMIC THEORY – PRICE, PRODUCER SURPLUS AND CONSUMER SURPLUS

This box introduces some basic economic theory that describes the rationale for non-market values. There are three concepts of value that can be estimated by the different types of valuation methods: price, producer surplus and consumer surplus. Each has associated problems.

Prices and quantities for many goods and services are determined in the market-place. Markets occur where the intentions of buyers on the demand side come together with the intentions of the sellers on the supply side. The quantity supplied is determined primarily by the costs of producing the goods, while the demand is derived from the well-being or utility which consumers gain from buying the goods. Market supply and demand are simply the quantities of the resource that will be supplied and demanded at various price levels (see Figure 4.2). The intersection point of the supply and the demand lines define the equilibrium price and quantity of the good.

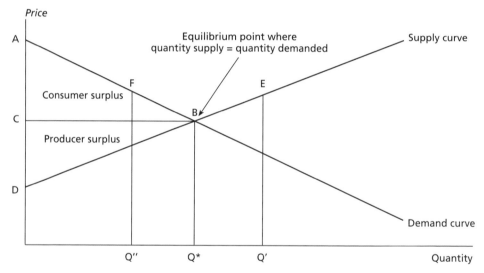

Notes:
Consumer surplus measures how much better off individuals are in aggregate when they are able to buy a good in the market. Each consumer values a good differently and as a result is willing to pay different amounts for goods. The demand curve is the summation of all these demands. The net benefit to consumers is measured by the area under the demand curve above the line representing the market price of the good (ABC).
Producer surplus is the sum over all units of production of the difference between the cost of production of the good, depicted by the supply curve, and the market price of the good (BCD). This is the benefit the firm receives.
The *maximum benefit*, that is the total producer and consumer surpluses, occurs where demand equals supply at point B. This equilibrium also determines the market price.

Figure 4.2 *Consumer and producer surplus*

This theoretical equilibrium has two interesting features associated with it. Firstly, the equilibrium price is a compromise value in that there are consumers who would be willing to pay more for the good. These consumers receive *consumer surplus*. Consumer surplus is the difference between what a consumer is willing to pay for a good and what he or she actually pays when buying it. The consumer surplus measure represents the benefits the consumer gains, represented by the area below the demand curve and above the market price line. It is denoted by the area *ABC* in Figure 4.2. On the production side, there are suppliers of the good who are willing to supply the good at lower prices than they actually do. These producers receive *producer surplus*. This surplus is the difference between the market price that the producer receives and the marginal cost to produce each unit. The producer surplus is the area above the supply curve, which represents the costs of production, and the area below the price line, which represents the value received by the seller for the good. It is denoted by the area *CBD* in Figure 4.2. The net benefits created by the market are the combination of these two surplus values.

The net benefits are maximized at the equilibrium point where the quantity supplied is equal to the quantity demanded. In other words, producing more or less of a good or service creates fewer benefits for society than is depicted by the sum of the producer and consumer surpluses shown in Figure 4.2. For example, if production is at level Q', then the cost of the last unit produced at point E would be greater than the price received. On the other hand, if the consumers only purchased at Q'', then consumers will still gain benefits from purchasing more because their WTP for their last unit purchased, depicted by point F on the demand curve, is greater than the market price. Given this result, the approach to valuing goods and services from an economic perspective is relatively straightforward for those goods traded in the marketplace.

Thus, in this theoretical framework, markets determine the prices for goods that maximize the net benefits to society. The equilibrium price is a representation of the value to society derived from an additional unit of a good or service being produced by a market. In projects that are relatively small in relation to the market, the market price is frequently used as the value of each unit of good or service created by the project, because the quantity produced will not affect the market price. Accordingly, these prices can be used in a cost-benefit framework to evaluate resource development projects. However, if a proposed project is large enough to influence the market, the prevailing market price is not representative of the extra goods and services being produced. Accordingly, changes in prices resulting from the increase in quantity supplied must be estimated.

This theory, of course, works well when markets exist. But in many cases markets do not exist and the goods and services of interest can be public goods or, in some cases, they can be quasi-private goods. These cases describe situations where individuals cannot hold property rights – or where they can, these rights are incomplete. For example, public goods have two unique features: non-rivalry where consumption of the good by one individual does not prevent others from consuming it – and non-exclusiveness – where it is difficult (or in some cases impossible) to exclude others from consuming the good. Examples of these involve air quality and national defence. Private property rights, on the other hand, characterize quasi-public goods in that they are exclusive. However, they are not freely traded in markets because of peculiarities with their supply. Examples of this class of goods include recreation in parks and the use of public libraries.

Even though markets fail with these types of goods, there are still issues related to their demand and supply. In simplistic terms, non-market valuation involves the estimation of consumer or producer surplus when these kinds of goods are at stake. Thus, the challenge of non-market valuation is to get individual consumers to reveal their preferences for public and quasi-private goods and services and to use this information to generate surrogate market prices that can be compared to those of private goods, for which the economic market functions well.

theory that underlies non-market valuation. The following section describes values and welfare measures, including concepts such as willingness to pay (WTP) and total economic values. The remainder of the chapter is dedicated to describing valuation methods, first-stated preference methods (contingent valuation) (see 'Stated preference methods: contingent valuation') and revealed preference methods (travel cost, hedonic pricing and related goods; see 'Revealed preference methods'). These are also illustrated through examples provided in boxes. Again, the purpose of this chapter is to describe some of these methods in general terms and to make readers aware of the basic differences between valuation approaches. The citations in this chapter refer readers to more specific information on the methods outlined.

Values and measures of welfare

Non-market valuation procedures typically generate monetary values associated with the good or service at stake. These values, commonly called welfare measures by economists, are frequently the subject of controversy. For example, some individuals believe that how they feel about non-market goods and services derived from forests should not, or cannot, be quantified in monetary terms. Monetary currencies are frequently used in non-market valuation because there is an imperative to create a common denominator that may be used to value a broad range of goods. Some studies have used other common denominators in order to aggregate values of a range of goods. Money is often used, not with the intent of creating markets for non-market goods, or for commodifying goods and services, but rather to create a simple, common metric for measuring a diverse set of values. Others are sceptical about the validity of the methods, or are unclear as to whether this information would be relevant in making public decisions. Much of this controversy involves misunderstanding the values that are generated and the claims made that these values are valid measures of human welfare.

Economic theory suggests that welfare can be assessed by implicit or explicit changes in prices and also by changes in the qualities or characteristics of goods and services. Thus, economic valuation is an anthropocentric process that does not value the environment per se, but assesses the preferences individual consumers hold for a particular change in some specified level of environmental quality. Underpinning this process is the concept that individuals are willing to make trade-offs among substitute goods and/or between money income and goods in order to make themselves better off. The valuation process considers trade-offs associated with changes in environmental quality.

Money measures of value indicate the rate at which an individual is willing to trade units of goods and services for money income. Thus, most of the methods typically measure the maximum amount an individual is willing to pay for the good or service under examination. This WTP is the payment required to make a consumer indifferent to the situation before a change in price or quality and after the change.[1] This idea describes consumer surplus (see Box 4.1). However, the payments required to make consumers indifferent to changes in prices or qualities can be evaluated from positions either before or after a change in the environment occurs. The payment required to make a person indifferent before a change occurs is called the *compensating variation* (CV). The payment required to make a person indifferent after a change has occurred is called the *equivalent variation* (EV). The selection of the appropriate welfare measure is important. For example, if a change could increase an individual's well-being (eg a development project), the maximum payment that person could make and still be as well off as they were before the change would be the CV. In the case of EV, the minimum payment the individual would accept to forego the change measures the welfare of the project after it has occurred. However, if the change would decrease the individual's well-being (eg a project which negatively affects the environment), the CV would be the minimum amount required to be paid to the individual to make them as well off as they were before the change. The EV would be the maximum they would be willing to pay to avoid the negative effects of the change.

This discussion may seem rather confusing; but the reason it is relevant is the fact that the demand curves (Box 4.1) that analysts observe (either market or the estimated non-market demands) are characterized by income effects. As individuals move along their demand curves with changes in prices and quantities, their levels of income also change. For example, as the price of the good in Figure 4.2 goes up from C to a level less than A, the quantity consumed declines below Q^*. The consumer may start to utilize substitutes (the substitution effect); but as the price goes up the consumer also becomes poorer because his/her income may not go as far relative to the situation before the

change. This result is called the income effect. The individual may consume less of the good in question as a result of utilizing substitutes, but also as a result of being somewhat poorer. Thus, consumer surplus, the welfare measure derived from observed or ordinary demands, may be biased due to the inclusion of this income effect. For CV and EV welfare measures, however, some measure of welfare is held constant and the income effect does not appear. Thus, CV and EV are the theoretically correct welfare measures.

The existence of income effects may appear to be a serious flaw in the use of consumer surplus as a welfare measure. Nevertheless, Willig (1976) showed that in most cases income effects are so small that the difference between CV/EV and consumer surplus is not significant. Willig's demonstration allowed economists to argue that consumer surplus was theoretically correct, particularly in the majority of published cases largely involving environmental impacts in developed countries. However, one can envision cases where income effects may be high enough to question the use of consumer surplus. These include cases where changes involve large-scale environmental damage or total destruction of some natural resource (eg a forest). Income effects may also be significant in developing countries where small changes in the environment may be associated with large income effects. Recently, advances have been made in the techniques available to measure welfare. Most of these techniques allow the direct estimation of CV or EV as a function of income (see Mitchell and Carson, 1989), as opposed to relying on inferring consumer surplus through estimating demands. Thus, income effects may not be as serious a concern as they once were.

The final issue to be discussed is: what exactly do non-market valuation methods measure? So far this chapter has addressed the existence and definition of welfare measures – but welfare measures for what? This topic has generated a great deal of discussion in the economic literature and what emerges is the idea of *total economic value*. Total economic value is the monetary measure of a change in an individual's welfare resulting from changes in the environment. This value can be decomposed into a number of components: use value and passive or non-use value.[2] Use values are derived from the actual consumption of elements of the environment or resource, or the in situ consumption of market goods that are complementary to this environmental resource. An example of a use value is outdoor recreational activities that involve the unpriced use of a forest. Passive use values do not involve consumption and essentially are held by individuals independently of their current use of those resources. Passive use values include existence value and bequest value. The concept of passive use value was first introduced by John Krutilla (1967), who argued that individuals do not have to be consumers of a resource in order to derive value from the existence of this resource. He wrote that 'when the existence of a grand scenic wonder or a unique and fragile ecosystem is involved, its preservation and continued availability are a significant part of the real income of many individuals' (Krutilla, 1967, p779).

Eliciting these values from individuals can be challenging. There are two sources of information from which values can be estimated. The first is observations of actual behaviour of individuals in real-world settings, such as markets. However, for non-market goods and services, markets can be simulated or constructed and values can be inferred from individuals' responses to hypothetical price or quality changes in these markets. These techniques are called direct methods because they elicit values directly from the respondent. Because the constructed market is hypothetical, direct methods require individuals to 'state' their preferences for non-market goods. Thus, these methods are also called stated preference methods.

The second set of approaches involves inferences of values through indirect measurement based on models of individual choice behaviour. These indirect methods require actual observations of individual choices and involve consumers revealing their preferences for non-market goods. As a result, these methods are classed revealed preference methods. While the stated preference methods can be used to estimate both use and passive use values, revealed preference methods require observations of human behaviour so that they can only measure use values.

Stated preference methods: contingent valuation

The principle stated preference (SP) method is contingent valuation (CVM). This approach involves the investigator collecting information directly from individuals using surveys that can be administered by personal individual interviews, group meetings, mail surveys or telephone surveys. In most applications respondents are asked to provide their WTP (or willingness to accept compensation) for specified increases or decreases in the quantity or quality of a non-market

---

BOX 4.2 SCENARIO DESIGN AND CONTINGENT-VALUATION MEASUREMENT OUTCOMES

Reference level of utility

1 The respondent should understand that he or she is currently paying for a given level of supply. In other words, the respondent should be aware of the property right.

2 The valuation scenario should clearly indicate whether the levels of the good being valued are improvements over the status quo or potential declines in the absence of sufficient payments.

3 The respondent should take his or her taxes and fixed long-term obligations into account in giving WTP amount for the good. In other words, the respondent should be aware of his or her current disposable income.

Nature of the public good

4 The nature of the good and the changes to be valued must be specified in detail.

5 Ensure there is no assumption on the part of the respondent that one or more improvements are included in the change that he or she is valuing when that is not the case (and vice versa).

Relevant prices of other goods

6 Where the change in the public good will significantly affect the prices of other goods, the impact of this change should be communicated to the respondent.

Conditions for providing the good and its payment

7 A respondent should be made aware of the frequency of payments required (monthly, annual, etc) for the quantity or quality change in the non-market good and whether the payments will be required over time in order to maintain the change.

8 The CVM scenario must also include who else will have access to the new levels of the good and who else will pay for it.

Nature of the WTP amount desired

9 The scenario design should express the consumer's surplus for the good and not some other notion of value, such as the 'fair price' (see Box 4.1).

Source: adapted from Mitchell and Carson, 1989

---

good or service. Values provided by this method are hypothetical in that the respondent is assumed to behave as if there were a real market. Hence, the valuation is contingent upon the creation of this hypothetical market.

Much of the difficulty and controversy with the method is related to the validity of this hypothetical market. Mitchell and Carson (1989) provide what is probably the most thorough treatment of CVM. They discuss five areas (pp50–52) which a CVM study must define in order to successfully establish this hypothetical market. These guidelines are summarized as points under their five headings in Box 4.2. The key to ensuring that these conditions are met in a CVM study is the design of the questionnaire and the specific administration of the CVM question(s). Successful CVM questionnaires include four components. The first is a series of 'warm-up' questions that place a respondent in an appropriate frame of mind.

The second is the description of a hypothetical scenario in which the market is established (see Box 4.3 for some examples). This description is one of the critical aspects of the study and must set out the terms under which the quantity or quality of the good will be provided. The current levels of provision are outlined (the reference level of utility), the quality and reliability of provision, the method and frequency of payment, and related information on the constructed market.

The third component involves the actual valuation questions which require a respondent to determine how much she or he would value a good if confronted with the terms and conditions specified in the scenario presented previously in the survey instrument. In some applications, respondents are reminded of their income levels and/or the requirement for them to make adjustments in their expenditure patterns. This is due to the existence of substitute goods or reductions in the purchase of other goods as a result of their reported increase in payments for the non-market good at issue.

Finally, the fourth component involves the gathering of information about the socio-economic or demographic characteristics of the respondent. This information is used to examine individual influences that may explain the respondent's WTP, and can also be used to aggregate individual responses to some policy-relevant aggregate population.

Elicitation methods

There are a number of ways that values can be elicited from respondents using CVM questions. The various elicitation methods, examples of which are shown in Table 4.1, have evolved over time, and new methods were usually developed in response to a specific theoretical concern or the refinement of econometric techniques. Early applications of the CVM simply asked a direct question, such as: 'What is the most you are willing to pay for the non-market good?' Or researchers enhanced this with iterative bidding games in which the interviewer raised or lowered bids from some arbitrary starting value until a respondent switched their reaction from rejection to acceptance (or vice versa) of the bid amount. These methods are now called open-ended CVM (Table 4.1). Gunawardena et al (1999) and Boadu (1992) implemented these questions in CVM studies in developing countries. However, concern with implausibly high or low bids in the direct question, or with bias associated with the opening bid in the bidding extension of this method (eg Randall et al, 1974), resulted in the development of alternative elicitation methods.

The payment card method (Table 4.1) extends the open-ended approach by providing respondents with an array of dollar values starting at \$0. The individual is asked to choose one of the numbers in the list, or some value in between two of the numbers, that indicates the maximum WTP for the good. One of the first applications of this method was by Mitchell and Carson (1984), who were concerned about starting point bias with the bidding game approach. However, they admit (Mitchell and Carson, 1989, p242) that the use of payment cards to capture values may introduce bias as a result of the range of values presented in the question. This apparent 'range bias' is largely untested (Rowe et al, 1996), but the possibility of this problem may have resulted in few applications of this elicitation method.

BOX 4.3 THREE EXAMPLES OF ACTUAL QUESTIONS USED IN CONTINGENT VALUATION STUDIES IN DEVELOPING COUNTRIES

Example 1: Madagascar (Kramer et al, 1994)

Suppose you are asked to use only the buffer zone set aside for collecting forest products and for growing crops, and are asked not to use the rest of the forests any more. Suppose, in order to make up for asking you not to use the forests in the park, you are given ___ vata[*] of rice every year from now on. Would this make you as content as before when you could use the forest in the national park?

Example 2: Nigeria (Whittington et al, 1992)

a   If this monthly fee were 5 naira[**] per month, would you be willing to pay this fee or would you choose to continue to get water from springs and tankers?

YES – pay the flat fee of 5 naira per month   if YES go to (b)

NO – get water from tankers or springs   if NO go to (c)

NOT SURE   if NOT SURE go to (c)

b   If this monthly fee were 15 naira per month, would you be willing to pay this fee or would you choose to continue to get water from springs and tankers?

YES – pay the flat fee of 15 naira per month   if YES go to (c)

NO – get water from tankers or springs   if NO go to (c)

NOT SURE   if NOT SURE go to (c)

c   What is the most you could afford to pay each month for water from public taps? _____ naira per month.

Example 3: Indonesia (Kramer et al, 1997)

QUESTIONNAIRE FOR FARMING HOUSEHOLDS IN VILLAGES AROUND RUTENG NATURE RECREATION PARK

I will be asking you a number of questions about the Ruteng Nature Tourism Park. This park is established by a group made up of Forest Protection and Nature Conservation (Perlindungan Hutan dan Pelestarian Alam or PHPA), other government agencies and many local non-governmental organizations (NGOs). The park managers would like to help and therefore want the opinion of the local people. The purpose of the park is to protect the forests, wildlife and the environment on Flores, and to provide new economic opportunities for the people living around the park.

Have you heard about this park?   1 = Yes   2 = No

Have park officials/staff ever visited you?   1 = Yes   2 = No

Deforestation is probably the main reason for a decline in dry season streamflow in the Manggarai areas. The Ruteng Park plans several activities: protecting existing forests, planting trees in degraded watersheds and teaching the farmers new soil conservation measures. These activities will provide drought control services, which involve decreasing the drought conditions for crops and improving a supply of dry season water. Suppose all those who live and farm near the park had to pay an annual contribution to the Taman Wisata Alam Ruteng Park to cover its annual costs. Assume that the Ruteng Park successfully implemented its watershed protection activities.

a   Will you be willing to pay an annual fee of _____ rupiah*** for this drought control service provided by the park?

Yes = 1   No = 2

b   If yes to (a), will you be willing to pay an annual fee of ____ (X + Y) ____ rupiah for this service?

Yes = 1   No = 2

c   If no to (a), will you be willing to pay an annual fee of ____ (X – Y) ____ rupiah for this service?

Yes = 1   No = 2

d   If no to (c), what is the maximum you could
afford to pay each year for the drought mitigation service   _____ rupiah?

e   If (d) is zero, why do you not want to pay for this service?

QUESTIONNAIRE FOR ECOTOURISTS

The government of Indonesia is establishing the Ruteng Nature Recreation Park, which includes Golo Lusang, Lake Ranamese, and Mount Pocoranaka. The forest in the park contains unique species of plants, birds and animals. The forest also protects rivers and streams that provide water for nearby villages and farms. The main purpose of the park is to protect the forest. The park will eventually contain a visitors' centre, picnicking and camping facilities and hiking trails so that visitors may observe forests, wildlife and high mountain views.

Now imagine that the park facilities are already available.

How much more time would you have spent in the Ruteng region if these facilities were available?

____ none

____ one more day

____ several more days

____ other (please specify) _____

Suppose that the park charged each visitor an entrance fee (a single fee for each day) and the revenues were used locally by the park to protect the forest and assist nearby communities. Considering what you have already paid for your trip to Ruteng, what is the maximum additional fee you would be willing to pay per person to visit the park?

US$____ or rupiah _____ rupiah per person

* A local unit of rice equalling 30kg. ** US\$1=20 naira. *** US\$1=2300 rupiah.

Table 4.1 *Examples of question formats in contingent valuation studies: measurement of a change in quality from Q_1 to Q_2 for a hypothetical forestry activity*

Type of question format	Example
Open ended	What would be the maximum amount you would be willing to pay to change conditions for collecting mushrooms from Q_1 to Q_2?
	$ _____
Payment card	Which amount below is the amount that you would be willing to pay to change conditions for collecting mushrooms from Q_1 to Q_2 (please circle your response)?
	$0 1 2 3 4 5 6 7 8 9 10 15 20 25 30 35 40 45 50 55
Closed ended	Would you be willing to pay $(bid)* to change conditions for collecting mushrooms from Q_1 to Q_2?
	Yes ☐ No ☐
Closed-ended referendum approach	If you had to vote for a programme that would change the conditions for collecting mushrooms from Q_1 to Q_2, and it would cost your household $(bid)* more in levies and taxes each year, how would you vote?
	For ☐ Against ☐

* A specific amount is put in the questionnaire, and this amount varies among questionnaires.

The most popular elicitation method is now the discrete response. This procedure, also called the closed-ended method (Table 4.1), utilizes a take-it-or-leave-it approach to the valuation question. In essence, the question asks a respondent to accept or reject a specific monetary amount that is drawn from a previously devised distribution of values. Thus, each respondent may face the decision to pay a different amount, and a regression model can be estimated that predicts the probability of accepting the environmental change as a function of individual characteristics (eg income) and price. The mean price from the estimated regression model represents the average WTP.

The discrete response method has several key advantages. The first is that the question mimics a market setting where the price is stated by the seller and the individual faces a choice of accepting the price and purchasing the item or not. A second advantage is that the method is easier for individuals to answer, making the procedure easy to implement in survey settings. In contrast to the bidding game and payment card methods, the respondent simply makes a single decision about the given price and other information provided in the survey.

One concern with this approach is that individuals may not have an incentive to truly reveal their WTP. For example, a respondent may be willing to pay the amount offered, but would reject the amount, thinking he or she could free-ride on the payment of others. In response to this concern about 'incentive compatibility', economists devised the referendum method; this involves presenting the discrete response question in the form of a referendum (see Table 4.1) in which all citizens would vote. The idea behind the approach is to reduce the chance of individuals free-riding: if the hypothetical proposal were accepted through a referendum, then all citizens would have to pay for the proposal regardless of their individual vote.

While the discrete response method is currently the method of choice for CVM practitioners, there have been extensions of the method that provide more information about the WTP of respondents. The principle advance has been the inclusion of follow-up questions that provide a higher or lower WTP amount, depending upon whether the respondent accepted or rejected the first amount presented. This approach responds to the criticism that the discrete response only provides an answer to the single question: 'Are you willing to pay $X for this good or quality change?' Thus, the discrete response method does not provide an estimate of the maximum WTP of each individual respondent. The addition of further WTP questions that are dependent upon the respondent's first answer improves this information base and is called 'bounded' CVM. Research suggests that only one or two extra questions should be asked, and that the extra questions should use prices that are substantially different than the original amount.

Collecting CVM data

Since CVM studies involve the researcher providing information to respondents, surveys must be used to collect CVM data. The design of successful surveys requires considerable skill. Two aspects of this design involve selecting the sample population to whom the survey will be administered and developing the survey instrument or questionnaire. The sample must be drawn such that the results will represent the population relevant to the issue under investigation. For example, if the non-market values associated with a fishery are of interest, then samples may have to be drawn to represent subsistence users and recreational users. The strategies used to design appropriate samples can be complex. Those interested are encouraged

to examine the statistical survey literature (eg Dillman, 1978).

The design of the survey instrument involves a number of considerations, the first of which is probably the method of administrating the questionnaire to the sample. This can involve in-person interviews, telephone or mail. Each of these approaches has strengths and weaknesses. The in-person interview is costly, while the telephone and mail methods involve much lower costs. On the other hand, the in-person interview can generate high-quality information due to the fact that interviewers can motivate respondents and provide detailed explanations of the complex nature of CVM scenarios. This allows for more informed respondents and greater accuracy of interpretation of the study. The method also allows for the presentation of visual or narrative aids that can be used to explain environmental impacts more fully to subjects. These features are possible to a certain extent in mail surveys, but not in telephone surveys. Mail and telephone surveys may be particularly inappropriate in developing countries, given the problems with telecommunication systems and the greater accessibility of such systems to a relatively small elite within the country.

Mail and telephone surveys can be effective if the tools developed by survey researchers are used. Dillman (1978) describes many of these tools. However, if the survey has not been designed and implemented correctly, the information collected may suffer from non-response bias. This can be examined to a certain degree (Mitchell and Carson, 1989); but one is never sure if the individuals who did not respond to the survey represent some systematic component of the relevant population. Many researchers examine demographic characteristics of the respondents and compare them to the sample target population. However, demographic

Table 4.2 *Scenario design and contingent valuation measurement outcomes*

Is the scenario ...	If not, the respondent will ...	Measurement consequence
Theoretically accurate?	Value wrong thing (theoretical mis-specification)	Measure wrong thing
Policy relevant?	Value wrong thing (policy mis-specification)	Measure wrong thing
Understandable by the respondent as intended?	Value wrong thing (conceptual mis-specification)	Measure wrong thing
Plausible to the respondent?	Substitute another condition or	Measure wrong thing
	Not take seriously	Unreliable, biased, don't know or protest zero response
Meaningful to the respondent?	Not take seriously	Unreliable, biased, don't know or protest zero response

Source: adapted from Carson, 1991

features may not capture characteristics such as attitudes or participation levels in various activities or organizations. CVM practitioners attempt to achieve high response rates in their surveys so they can argue that non-response bias is negligible. In some cases, researchers with fixed budgets are best advised to reduce sample sizes and increase funds spent on tactics in order to generate higher response rates. In general, response rates in the 80 to 90 per cent range are targeted.

Issues with CVM

CVM is a popular approach to measuring non-market values but is still considered controversial. Carson (1991) points out that the market created in the questionnaire must meet requirements imposed by economic theory and that the respondent must fully understand the good being valued, as well as the market. He presents a set of criteria (see Table 4.2) that can be used to evaluate the design of the questionnaire. Many of these criteria, if not met, can result in a mis-specification of the scenario to the respondent. This can generate bias: the respondent may not be familiar with the good or its characteristics being valued. In addition, surveyed individuals may not take the questionnaire seriously and will not respond, or will respond inappropriately and misstate their value. Tests have been developed by researchers to examine whether the answers to CVM questions vary systematically from what would be expected according to economic theory (eg Whittington et al, 1990). In general, it is essential that researchers spend considerable time and effort designing the CVM scenarios in order to ensure that a reliable survey instrument results. Extensive pretests of the questionnaire prior to its administration to the sample can uncover many problems.

Another bias uncovered by critics of CVM is strategic bias. This bias involves respondents thinking that they can influence the policy or investment outcome decision by the manner in which they answer the CVM question. For example, if respondents feel that they would like the good provided in the question and would like someone else to pay for it, they may provide or accept an amount higher than they actually would in order to coerce decision-makers to provide the good. In this case the strategic bias involves respondents overstating their WTP. On the other hand, respondents may view the survey as an instrument to estimate prices to be charged for the provision of the good. In this case the respondent may understate the WTP. Despite the potential for these biases, research examining strategic bias in developed countries has generally not found evidence for its occurrence (Mitchell and Carson, 1989).

Two other problems identified by Kahneman and Knetsch (1992) involve order (or sequencing) and embedding effects. The order effect occurs when the values for a particular good differ between surveys because the good is in a different position in the sequence of each survey's questions. Kahneman and Knetsch found that the same good generates a higher WTP when it is the first good to be valued and a lower WTP if its value is elicited further down in a list of goods to be valued. Since the order of the goods was arbitrary, and resulted in differing valuations, it raised the question of the validity of the responses. The embedding effect can be illustrated by the following example. An individual's WTP to protect a specific endangered species may be $100. However, if the survey first asked what the individual was willing to pay to save all endangered species, and then asked his or her WTP for a particular endangered species, the WTP for the single species would be substantially lower than $100.

These issues have caused serious concern with CVM and have spawned a great deal of research into the underlying reasons for the effects and for methods to alleviate them (eg Carson and Mitchell, 1995). These issues, and the fact that CVM could be used to assess environmental and resource damages in the US, has spawned considerable debate into the validity of CVM. The interested reader is encouraged to examine the exchange between Hanemann (1994) and Diamond and Hausman (1994) for an overview of the concerns.

CVM in the developing world

Contingent valuation methods were created by economists in the developed world initially for application in the developed world. However, there are increasing applications of CVM in the developing world (Whittington, 1998; see Boxes 4.4 and 4.5). A number of specific challenges are involved in translating this valuation technique into an effective tool in the developing world, particularly in places with subsistence or semi-subsistence economies. It is important to recognize differences in cultural contexts between industrialized and traditional societies. These differences may be manifested in different preference structures among cultural groups. The CVM and other non-market valuation methods are highly dependent upon economic theory, since they involve the estimation of welfare and values based on assumptions about the form of individual, not group, preferences. Some of these assumptions may not hold and provide challenges to researchers. A number of these challenges are identified in Box 4.6.

Perhaps the most serious challenge to the use of conventional CVM is the issue of monetary values, currencies and the timing of payments. In many cases money may not be the appropriate vehicle to assess trade-offs because there may be alternative mediums of exchange; money incomes may be also extremely small or, in some economies, simply not important. CVM practitioners have developed some ingenious ways to overcome these obstacles by using other types of resources as mediums of exchange. For example, Kramer et al (1994), in a study of protected area policies in Madagascar, use rice instead of money because rice is a common instrument in barter transactions. Their CVM question (see Box 4.3) asks respondents to accept units of rice in

BOX 4.4 THE USE OF CONTINGENT VALUATION TO EXAMINE HOUSEHOLD DEMAND FOR IMPROVED SANITATION SERVICES IN GHANA

In many cities in developing countries sewerage systems have been constructed to which residents cannot afford to connect their homes. If the costs of connection are reduced and the financing of the sewerage system comes from monthly user charges levied by municipal governments to finance sanitation improvements, information on how households would respond to such fees and charges is often missing. This study used contingent valuation to gather such information from residents of Kumasi, Ghana (Whittington et al, 1993).

The CVM questionnaire was developed after considerable experimentation and pre-testing. The authors used information from 50 household interviews and a number of small group discussions to design a questionnaire. They then pre-tested the survey with a sample of 100 households. The final questionnaire collected information about the household – in particular, its existing use of water and sanitation facilities, percep-tions of cleanliness, privacy and convenience, and its current monthly expenditures. The contingent valuation question examined the household's WTP for improvements to its current sanitation facilities.

The study collected information through personal interviews of 1224 households. The interviews were conducted in the predominant local language and the interview-ers were trained in how to ask questions and elicit answers. Interviewers described relevant sanitation options by reading from a prepared text and showed diagrams and pictures to respondents. The study examined respondents' WTP for five different sanita-tion services: Kumasi ventilated pit latrines; toilets (water closets) with sewer connections; sewer connections for already existing toilets using septic tanks; private water connections; and for households without water connections, their WTP for toilets with sewer connections. Households were asked for their WTP for only those options relevant to their current circumstance. The study examined the WTP of both landlords and tenants.

Respondent WTP was elicited through an abbreviated iterative bidding procedure (two questions) followed by an open-ended question. Thus, this approach involved a combination of yes/no questions and a direct open-ended question (see Box 4.3 for an example). The study also examined the sensitivity of WTP to the design of the question by adjusting the starting value of the initial bids. Some respondents received a high starting value and others a low starting value.

Multivariate models of WTP were developed from the completed interviews using four types of explanatory variables. Variation in the WTP for a given sanitation technol-ogy was found to be a function of the questionnaire's characteristics (whether a respondent was given a low or high starting bid), the respondent's characteristics (gender, education), the household's socio-economic characteristics, such as income, and the household's current level of water and sanitation services. A number of relationships between WTP and explanatory variables have clear economic interpreta-tions. For example, households with higher incomes were willing to pay more for improved services; owners of dwellings bid more for improvements than tenants; and respondents who were dissatisfied with the current service were willing to pay more to improve service. These relationships are similar to those commonly found in CVM studies in developed countries, lending support to the idea that CVM can be used in developing country contexts.

The results of the study suggest that most households were willing to pay more for an improvement in their current sanitation services. However, their values were not high, averaging about US$1.40 per household per month, or about 1 to 2 per cent of household income. The values did not differ appreciably for households who already had better sanitation systems. However, the demand for water and sanitation was found to be additive. Households without water connections reported WTP for US$1.52 per month for a water connection, and US$2.5 per month for both a water connection and a water closet.

These results supported the notion that conventional sewage systems are probably not affordable to most of the households in Kumasi without large government subsidies. However, the authors discovered that respondents were willing to pay as much for a ventilated pit latrine as they were for a water closet. Providing latrines is less expensive than installing toilets, and the authors suggest that if households could engage in financial transactions under terms similar to those in industrialized countries, household WTP for improved sanitation would be sufficient to pay the full costs of providing latrines.

Note: See Boxes 3.8, 4.5 and 4.8.
Photograph by Neil Byron

Figure 4.3 *A scene in Mana Pools National Park in Zimbabwe. Throughout Zimbabwe, revenues from wildlife tourism are being returned to the authority managing the resource, be it the wildlife department, local communities or landholders. The question of how to price the services provided needs to be addressed by means of the approaches outlined in this chapter*

BOX 4.5 THE USE OF THE CVM TO ESTIMATE NATIONAL PARK ENTRANCE FEES: AN EXAMPLE FROM COSTA RICA

How much should be charged to enter a protected area? Should residents and foreigners be charged the same? Can non-market valuation techniques such as the CVM be used to estimate optimal entrance fees in a developing country setting? In many developing countries, national parks are important sources of direct and indirect revenues and are necessary to protect biodiversity and national heritage. The use of entrance fees for national parks and protected areas is justified in order to:

- Generate revenues to recover costs and to ensure quality goods and services.
- Reduce congestion in overcrowded parks while promoting visitation in less crowded parks through differentiated pricing.
- Remove subsidized competition with privately owned protected areas.
- Promote equity by having direct users pay for parks, while foreign visitors pay higher fees than residents who themselves contribute to the establishment and maintenance of parks through general fund taxes or foregone opportunity costs.

A 'dichotomous-choice' CVM survey was recently used by Schultz et al (1998) to elicit WTP among Costa Rican and foreigners for future (repeat) visits to two of the country's most popular national parks (the Poas Volcano and the Manuel Antonio Beach Preserve). Resulting WTP estimates were expected to assist in evaluating the potential trade-offs between higher entrance fees and reduced levels of visitation.

This research was initiated because, between 1994 and 1996, Costa Rica changed national park entrance fees three times without using any formal or objective economic analyses. This resulted in much controversy and public debate. Current fees at the time of the study were not considered optimal in terms of revenue generation or as an effective strategy to equalize visitation among different parks.

On-site visitors to the parks were asked about their backgrounds and perceptions of the parks, including the following CVM question:

If the infrastructure and services in this park are
greatly improved, would you be willing to pay $(bid)
for the entrance fee in a future visit (yes/no)?

Usually CVM questions include more details concerning the costs that respondents are asked to pay in order to avoid 'information bias'. Here, the specific details about the park's infrastructure and services were intentionally left out of the WTP question for two reasons. Firstly, these details were omitted in order to compare WTP values for two parks with very different types and levels of infrastructure and services. The Poas Volcano is very developed while the Manuel Antonio Beach area is rustic. Secondly, the survey focused on visitors exiting the parks who were assumed to be familiar with park infrastructure and services. The resulting values are displayed in Table 4.3.

These mean WTP estimates indicate that higher park entrance fees should be charged in order to improve the infrastructure and services of the parks. Foreign visitors value the volcano more than the beach, while the opposite is true among local visitors. This is likely because there are many beaches in the homelands of the US and European visitors but few active volcanoes. In the Central Valley of Costa Rica (where most residents live), there are many volcanoes and few beaches. Finally, logistic WTP curves, such as those shown in Figure 4.4, can be used to predict changes in WTP (and visitation) associated with proposed changes to park entrance fees.

Table 4.3 *Mean WTP for repeat park visits with improved infrastructure and services*

	Poas Volcano (US$)	Comparison to actual fees (%)	Manuel Antonio Beach (US$)	Comparison to actual fees (%)
Foreign visitors	23	100	14	100
Local residents	11	+800	13	+900

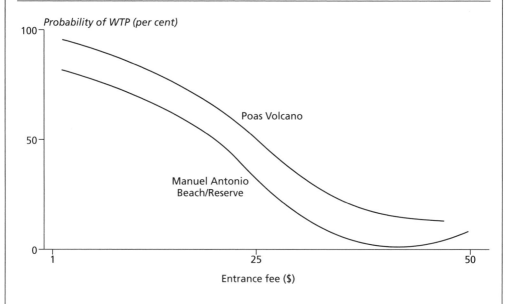

Figure 4.4 *Foreigners' WTP curves for future visits to two Costa Rican national parks*

return for not using forested areas for collecting forest products.

The question of whether to use money or some other commodity as a medium of exchange relates to an issue economists call the payment vehicle. This term, however, relates not only to the medium of exchange, but also to the manner in which the payment is collected. For example, one could use the notion of donations to a hypothetical fund or potential taxes to be collected by governments. In some cases, however, weak government institutions may undermine the credibility

of some types of CVM scenarios. This may be germane in developing countries. Accordingly, practitioners should pay careful attention to the plausibility of payment vehicle being used in a CVM scenario.

Boadu (1992) encountered problems with the timing of payments in his study of WTP for potable water in rural Ghana. Because of the way local markets operated, Boadu used the iterative bidding approach in the CVM question and used monthly payments as a payment vehicle. However, he found that monthly payments

---

BOX 4.6 KEY CHALLENGES TO NON-MARKET VALUATION IN THE DEVELOPING WORLD

Lack of substitutability between goods

This is perhaps the most critical potential problem as most non-market valuation techniques hinge on the assumption that a respondent is willing to exchange a given good in question with another good. In the West, socio-cultural traditions around natural resource use have been largely utilitarian. More recently, recreational use of natural resources has become important and therefore more highly valued. In many cultural contexts in the developing world, natural resources are embedded in the cosmology of local cultures. Many more natural objects and places are considered sacred or taboo, and sacred or taboo goods are often either non-substitutable, or there may be greater constraints on how they are used or whether or not they may be exchanged. For example, in Zimbabwe some individuals believe that it is inappropriate for certain fruits to be traded in markets. It may be the case that a higher proportion of natural resources falls into the category of non-substitutable goods, rendering valuation of these goods impossible in quantitative terms. At the least, such conditions create challenges for making accurate estimates or translating them into currencies. One can still recognize and describe sacred and taboo valued goods, but perhaps not in a comparable context.

Property rights

Most non-market valuation techniques embody certain assumptions about property rights. This is particularly true of direct methods that attempt to elicit non-market values by presenting respondents with scenarios involving hypothetical property rights. Respondents must be able to imagine that the proposed change could actually take place. In addition, they must be able to imagine how such a change would alter their individual utility. And finally, they must understand how that utility change could be measured in some common unit of currency. Any numbers of problems can arise. For example, if grazing land is communally accessed, it may be impossible for an individual to place a value on his or her relinquishment of those rights. The value ascribed to that right would be contingent upon whether or not everyone else in the collectivity was similarly restricted. An individual's property rights are linked to communally held property rights, and instruments intended to measure change in individual utility would have to account for change in social welfare derived from communal rights.

Individual versus group sovereignty

Sovereignty issues have to do with the ability or authority to make decisions. This issue is therefore closely connected to the property right issue. When, as in the above example, resources are held collectively, decisions regarding their allocation and use are often made collectively. This may make it difficult for individuals to value goods which they did not own or to which they did not have exclusive access.

Satiation

Many subsistence goods have limitations in the degree to which they may be accumulated. If there is no market, or no ability to use or benefit from surplus production, marginal returns to producing additional units of the good may be zero. There may also be cultural norms and traditions regarding natural resources that place positive values on the consumption of a good up to a certain point (culturally sanctioned levels of consumption) and negative values on consumption beyond the satiation point. An individual's status within the community may decline if he or she is perceived as being greedy or wasteful, and therefore values for a given good may vary dramatically depending upon the amount consumed.

Demographic effects on non-market values

The task of sampling populations of resource users in the developed world is often straightforward. Phone lists may be used for random sample surveys. User lists are available for many natural resource-use activities (boat licences, fishing licences, harvesting permits, etc). In the developing world and in subsistence or mixed economies, variables such as gender, generation and location may have critical effects on aggregate values for natural resources. Sampling may be very challenging if researchers' access to a broad cross-section of the population of interest is restricted. It may be difficult to determine who uses a given resource for what and how to reconcile such different values. Men and women might value trees very differently depending upon their location and their specific gender uses. Another problematic aggregation issue arises when attempting to articulate values concerning individuals or households with a broad range of incomes at large regional scales. Income levels are known to affect values, and the existence of great disparities may lead to distortions when attempting to aggregate values across income strata.

Source: Adamowicz et al, 1997

---

may not have been appropriate because farmers harvest at different points during the year. Furthermore, their wealth may not be monetary but in grain. This caused problems because farmers may not have cash income on hand each month from which to pay monthly water bills.

Lynam et al (1994) and Emerton (1996) have made some interesting modifications of CVM for the valuation of woodland resources in Zimbabwe and Kenya, respectively. Using cards that depict environmental goods, respondents in household questionnaires were asked to distribute a certain number of counters among the cards. One of the cards depicted a good for which a monetary value was known or for which a typical CVM question was asked. The values for

the various goods were then derived from the known good. While the methods undoubtedly require further development (see Chapter 6), they indicate some of the modifications that may be worth pursuing in developing country situations. The use of cards and counters is largely derived from techniques in participatory rural appraisal, which are further discussed in Chapter 6.

An exploratory application and assessment of CVM was undertaken in Zimbabwe to value various land uses (see Box 4.7; Adamowicz et al, 1997). This led us to the development of an alternate scenario using participatory research appraisal techniques in the group setting. It is suggested that the values derived should only be regarded as indicative.

BOX 4.7 CONTINGENT VALUATION: AN EXPLORATORY CASE STUDY
USING PARTICIPATORY RURAL APPRAISAL TECHNIQUES

An exploratory application and assessment of CVM was undertaken during a workshop in Zimbabwe with a multidisciplinary group. The objective was to value various land uses in the communal area adjacent to Mzola State Forest. One component of the workshop was to assess alternate uses of the woodland area. An option was to use the area for agricultural purposes, while another was to use it for commercial harvesting of indigenous trees, with concurrent commercial grazing of cattle. WTP procedures were initially explored using two scenarios. One was designed to elicit values of agricultural land. The other attempted to value communal forest lands. The procedure involved designing a questionnaire that was translated into Ndebele and back-translated to check accuracy. Interviewers were trained in extensive role-acting sessions and the questionnaire was considerably adapted during the process of pre-testing in the field. Closed-ended referendum value questions were used (see Table 4.1) and farmers were asked if they would be willing to accept the specified values per acre annually for the loss of specified areas of agricultural land. The specified areas and monetary amounts were varied in a random manner for different respondents.

The use of CVM to value agricultural land proved to be successful in that the respondents apparently understood the scenario and provided values that conformed to theoretical expectations. However, the use of CVM to value communal woodlands, which involved a scenario incorporating compensation to households individually and communally, was less credible and thus less successful in terms of eliciting valid responses from individuals. This led to the development of an alternate scenario using participatory rural appraisal (PRA) techniques in the group setting of a village meeting. Through this means, it was possible to assess the potential for co-management of an area of the communal woodland by villagers and a concessionaire. It was hypothesized that land-use conflicts might be alleviated by a co-management arrangement in which communal farmers might be paid by the concessionaire, who would conduct commercial timber extraction and replanting. The WTA of villagers, conditional on their acceptance of joint management and on their being paid for their participation in the co-management arrangement, was assessed by a process of open voting. It was expected that a 'village payment' per household to be paid directly to a village council (VIDCO) for communal purposes might be preferred. However, a considerable majority indicated a preference for monetary compensation to be paid directly to the households. The lack of statistical validation, and the potential problem of value-embedding for the values elicited by the exercise of a single village meeting, suggest that these should only be regarded as indicative.

Revealed preference methods

The other class of methods discussed above involves inferences about non-market values through indirect measurement. Successful application of these methods requires observations on actual choices or purchases made by individuals in real markets. In these cases, economists believe that this behaviour involves individuals revealing their preferences to researchers; thus, many economists call this class of methods revealed preference (RP). Although there are a number of these techniques, three of the most popular ones are described in the following sections. These include the travel cost model, hedonic pricing and related goods approaches.

The travel cost model

Consider the case where an analyst observes individual recreationists or tourists who choose to visit one national park but not another. In this case, these individuals have revealed their preference for a specific national park. In their selection of this site, the individuals had to incur travel costs in order to visit. Thus, the market purchase involves the payment of the travel expenses and entrance fees to visit this particular park. Each visitor (or group, if travelling together) had to face different costs to visit this park, as well as the park they chose not to visit. These travel costs can be considered implicit prices that reflect behaviour. They can be used in place of conventional market prices as the basis for estimating demand curves from which consumer surplus estimates are calculated (see Box 4.1). In essence, the consumer surplus represents the amount that a consumer would be willing to pay in the form of an entrance fee to visit the site.

This simple idea was first recognized by Harold Hotelling in a now famous letter to the US National Park Service. His idea was that visitors must incur costs to visit parks and that these costs vary in relation to the distances they travel between their homes and the park entrance gates. Because the costs of private goods (petrol, lodgings, vehicle repairs, etc) and the park visits (which are usually publicly provided) are weakly complementary,[3] Hotelling reasoned that travel costs could approximate true prices of trips to

the park. At different prices (ie distances), different numbers of trips will be made to the park and these price-quantity pairs identify the demand curve for the park. This simple idea has subsequently spawned a huge literature in recreation demand and also contributed to the development of a number of sophisticated econometric models.

The first travel cost models assumed that visitors were homogeneous in all respects except their location and their socio-economic characteristics. These assumptions allowed analysts to model 'aggregate' demand for sites. The aggregate units were usually some form of administrative geographic zones from which visitors would travel. Zones were also constructed by dividing the region around a site into areas of increasing travel distance (ie concentric rings). Thus, this form of the travel cost model became known as the zonal travel cost model. The unit of observation in this case is the origin zone, not an individual visitor. The analyst thus observes a visitation rate, usually number of trips per capita or the number of trips to a site, from a zone divided by the population of the zone. The model is estimated by regressing the visits per capita against travel costs and other variables that describe the characteristics of the zone's residents. In order to identify the demand curve, travel costs are increased so that estimated visits are equal to zero. Once the demand curve is identified, the non-market benefits of the site are estimated by calculating the area under the demand curve.

The earliest forms of this model were used to value single sites (eg Knetsch, 1964) or the collective benefits associated with certain types of activities over a number of sites (eg Brown et al, 1964). This latter approach, called the aggregate or pooled travel cost model, involves aggregating trips and travel costs over a number of sites and estimating one regression equation. The approach allows one to add site characteristics to the set of independent variables. This was the first model that permitted the examination of impacts of site characteristics on the demand for visits to sites.

These traditional forms of the travel cost model have important assumptions:

- All trips made to the sites in the model are determined at one time – for example, at the beginning of the season.
- All individuals spend an equal and fixed amount of time at the site.
- The main purpose of the trip is recreation.
- Individuals treat entrance fees and travel costs as identical.
- There is no utility gained in the actual travel to the site – in other words, individuals do not enjoy the act of travel to sites.
- There are no substitute sites available.

McConnell (1985), Smith (1989) and Fletcher et al (1990) summarize these assumptions and other issues in detail.

The zonal travel cost model was used until criticism by Brown and Nawas (1973) shifted attention to individual-level data rather than zonal or aggregate demand data. The underlying concern with the zonal model is that it is not based on a model of utility maximization (McConnell, 1985). The method involves the strong assumption that every individual in the zone has the same demand parameters; in essence, the model assumes that the estimated demand is generated by a representative consumer whose behaviour reflects average behaviour in the zonal population (Fletcher et al, 1990). Thus, to circumvent these concerns individual-level data would be preferred to zonal data; this circumvents the

assumption that everybody in the zone has the same preferences.

The individual-level travel cost model is now the most popular form of the model. In general, the estimation of this model proceeds by observing the visitation of individuals to sites. Each observation in the data is a single individual; thus, individual trips are regressed against their travel costs to the site and their individual characteristics. In order for this approach to work there must be sufficient variation in the number of trips people make to the site. The pooled approach to estimating this model involves the 'stacking' of individual-level data from the group of sites, including the site characteristics of interest to the set of independent variables in the regression model.

Most of the developments in the travel cost literature during the 1980s and 1990s involved ways to handle imperfect information from individual-level data. For example, one issue that arises from individual data is that not all potential visitors make positive numbers of trips to the site; some will take no trips. Thus, non-participation is an issue in travel cost modelling and a number of econometric techniques were developed to address this issue (eg Hellerstein, 1992). Another example is the fact that people take trip frequencies that have non-negative integer values (ie 0,1,2,3,4, etc, trips). This fact created a large literature on count-data modelling approaches to travel cost analysis, instead of the usual ordinary least-squares regression techniques (eg Smith, 1988).

One should note that developing travel cost models involves intensive data requirements. For example, trip data can be collected from on-site surveys of visitors, or from surveys of households within some defined market area for the site(s) of interest. With this form of data collection the analyst must be aware of all

BOX 4.8 THE RECREATIONAL VALUE OF WILDLIFE VIEWING AT LAKE NAKURU NATIONAL PARK, KENYA

Lake Nakuru National Park was established as a bird sanctuary in 1961 and is the only waterfowl habitat site in Kenya falling under the Ramsar convention. The park hosts 360 species of birds, including approximately 1.4 million flamingos (*Phoeniocopterus* spp). The flamingos attract large numbers of visitors to the park. Of concern were decreases in the numbers of flamingos, linked to increasing water pollution from agricultural activities and increasing urbanization. The challenge for economists was to estimate the recreational value of preserving the current flamingo population in the park. To do this Navrud and Mungatana (1994) used both contingent valuation and travel cost methods. The travel cost results are described below.

Analysis of park records identified that both residents of Kenya and non-resident international tourists visit the park. The Kenyans usually take trips where the park is the single destination. However, the international tourists are often on multiple-destination trips, of which the park is probably only one stop. For these reasons separate travel cost models were estimated for the two visitor groups.

Information to construct the models was collected in the park by randomly sampling 185 visitors and administering questions in-person in the park. Of the 185 visitors, 58 were Kenyans and 127 were foreign visitors. The interviews involved determining length of stay, percentage of time in the park viewing flamingos, location of residence, names of potential substitute sites, subjective ratings of park experiences, and socio-economic characteristics. For the foreigners, the model involved the calculation of recreational value per visitor day for the entire trip in Kenya, the recreational value of the visitor's stay in the park, and the value of the portion of time in the park viewing flamingos. For residents, the analysis involved estimation of the demand for visits to the park only. Travel cost models were developed using individual observations, but the dependent variable (number of visits) was adjusted to represent the rate of visitation per capita. The full specification involved the following regression model:

$$(T_i/(USERS_{ij}/POP_{ij})) = f\ (P,\ I,\ A,\ S,\ E,\ P',\ F,\ Q,\ I')$$

where T = number of trips by individual i, $USERS$ = number of sampled users from zone j, POP = population of zone j, P = travel costs (round trip transportation and time costs), I = personal annual income, A = age, S = gender, E = education level, P' = travel costs to a substitute site (calculated in same manner as P), F = preference for flamingo viewing, Q = quality of visit to the park, and I' = annual household income in excess of personal income. Various definitions of origin zones were attempted, as were linear, semi-logarithmic and loglinear functional forms.

The results suggested that the 127 foreign visitors received net benefits of their overall visit to Kenya of about US$200,000. Since their trip lasted on average 21.16 days and the park visit lasted 1.52 days, the benefits foreigners received attributable to the park visit can be valued at US$114 to US$120 each. The benefits of the resident visitors were valued at US$68 to US$85 per visitor to the park. Applying this information to the total visitation level of the park in 1991 (141,332 visitors, 31 per cent of whom were residents), the recreational value of Lake Nakuru National Park in that year was US$13.7 to US$15.1 million. Of this amount, US$5–5.5 million could be attributed to flamingo viewing.

The results suggest that flamingos form a major component of the value of visiting the national park and that in order to maintain the current visitation rate, measures to arrest the population decline of flamingos may be warranted. However, the large consumer surplus values raise the question of how much the Kenyan government can appropriate of this surplus (see Box 4.1 for further discussion on consumer surplus). The Kenyan Wildlife Service's revenue from entrance fees and royalties from businesses operating in the park is only about 5 to 10 per cent of the estimated non-market values. The travel cost results identified inelastic demand for foreign visitors and elastic demand by resident visitors. This suggests that differential entrance fees could be established – increased fees for foreigners and lower fees maintained for residents. The Kenyan Wildlife Service recognized this possibility and increased the fees for foreign visitors in 1993 by 310 per cent.

of the problems of survey research (recall bias, non-response bias, sampling issues, etc). Ideally, however, the best type of information for travel cost analysis involves the use of mandatory on-site registration or permits. Many parks or other managed recreation sites require visitors to provide written information on where they plan to go, where they come from and what they plan to do in the area while visiting. Frequently, this is provided at the time an entrance fee is paid. If this type of data is available, sampling issues are resolved – the analyst has a census of use. The only problem with this data is that information from non-participants is missing.

These data issues usually constrain the use of travel cost models in developing country applications. Many parks in these countries do not use permits and when they do, they do not maintain databases of them. Surveys may be difficult to implement due to inefficient mail systems or unknown addresses. Thus, in developing countries analysts must resort to the use of on-site surveys. This technique has its own difficulties, including the spatial and temporal application of visitor interviews. Nevertheless, a number of useful models have been developed from this approach (eg Navrud and Mungatana, 1994; see Box 4.8).

Perhaps the most vexing issue with the travel cost model is the estimation of travel costs. The costs of visiting a site consist of the entrance fee, the transportation cost and the opportunity costs of time spent at, and getting to, the site. This latter component, the value of travel time, is one of the most difficult components to assess. The assumption is that time spent travelling is not enjoyable and is time that could be spent doing something else. Hence, this time must be considered an opportunity cost. In developed countries, this time is usually valued at some fraction (usually a third to a quarter) of an individual's wage rate (see Cesario, 1976). The theory is that if someone is not travelling, they could be earning their wage. However, this is controversial in developed countries because maximum working hours, holidays, taxes, etc, must be considered. In developing nations use of the wage rate may be even more problematic. For example, in developing countries wages may be non-existent, or the opportunity cost may involve valuing time spent doing something else, such as collecting fuelwood or participating in cultural or spiritual rituals. Hatton-MacDonald (1998), in an examination of fuelwood collection, addressed this problem by using energy expenditure (calories) instead of monetary travel costs as the 'price'

BOX 4.9 CHOICE OF FUELWOOD COLLECTION SITES IN ZIMBABWE: VALUATION THROUGH BEHAVIOUR AND CALORIC EXPENDITURE

Wood is the primary source of household energy for the developing world (see Figure 4.5). However, there is increasing concern about the state of forests in developing countries, in part because of the depletion of forest resources for use as fuelwood. In addition, the depletion of forest resources is seen as a significant contribution to greenhouse gas emissions and several suggestions for climate change policies include reducing the use of forest resources for fuelwood. In this box the value of forests as a source for fuelwood is examined by constructing a behavioural model of fuelwood collection for villages in north-eastern Zimbabwe (Hatton-MacDonald, 1998).

Use fuelwood is usually thought of as one of a series of choices that households face in allocating labour effort or cash income. However, in the research area, the sale of fuelwood is largely prohibited on communally held land and the prohibition is well enforced through local social institutions, such as the *'sabhuku's* court'. Therefore, households are dependent upon collecting their own fuelwood. As a result, a behavioural choice approach is put forth as a way to model the site choice problem.

The choice of any particular location will involve the opportunity costs of the time and effort required for the journey and the potential benefits of the collection site. Understanding this choice problem is of interest because it identifies the trade-offs that local people make, and identifies the behaviour that they would employ in response to changing forest resource stocks and conditions. Fuelwood collection behaviour was observed, and a model was constructed that 'values' changes in forest resources in terms of caloric expenditures. While it may be unusual to use calories as a measure of opportunity costs, calories may be much more relevant to people in small communities in developing economies than monetary units; here, cash transactions are rare and physical labour for subsistence is prevalent. Under these circumstances, caloric expenditure may explain more behaviour in terms of fuelwood collection than does the value of time or other monetary constraints (it is rare for individuals to be hired as workers because there is usually sufficient family labour). If the opportunity cost of time is not well described by wage rates, the next best alternative may be to use a measure of effort, such as time, difficulty ratings or an estimate of caloric expenditures involved in a trip to a fuelwood collection site. If calories are used in estimating models of choice, then calories provide an alternative means of expressing the welfare losses that the household or community may experience due to closure of the site.

The choice of particular site i can be modelled as follows:

Probability of taking a trip to site $i = f$ { travel costs (calories), availability of the species for firewood, other site attributes}

The empirical results suggest that calories (reflecting distance and difficulty of the journey) and attributes of the site, such as the availability of good-quality fuelwood, are important factors in the choice of sites. The welfare simulations reinforce the importance of the spatial context of fuelwood shortages. Closing sites may have a relatively small effect on the community but a large effect on the well-being of particular households. For example, if a site such as Marirangwe Mountain (a frequently chosen collection site in the research area) were closed, the closure would cost households on average 22 calories per trip. However, the cost could be as high as 107 calories per trip for some households. When households in this area are making two to three trips a week, caloric expenditures on a day-to-day basis are of fundamental importance.

The welfare effects have broad policy implications. For governments considering site closure to protect forested areas, the increased caloric expenditures by women will be a significant but less visible cost for the local population. A government or non-governmental agency that is mindful of these welfare implications has a few options available to redress the situation. For instance, compensation might be provided through deliveries of staple commodities (or cash equivalents) to increase caloric consumption. However, agencies must be cognizant that careful targeting may be required because the customary allocation of food within the household may not benefit those most affected by the site closure.

The estimation results and the welfare effects may also be of interest to governments from the industrialized world. With recent attention to global warming, governments and industries are interested in the potential for carbon sequestration in the developing world. This research suggests the nature of the costs that would be borne by the local population if stocks of carbon in the form of forested areas were set aside for protection.

Photograph by Neil Byron

Figure 4.5 *Wood being transported to the market in Quang Tri, Vietnam. While in many areas wood is traded and can be valued using market-based techniques, in other areas it is part of the subsistence economy where travel cost methods may be appropriate*

variable (see Box 4.9). Hatton-MacDonald measured costs in terms of the time required to walk to the site, calories expended to reach the site and the perceived difficulty in accessing the site. As the costs increased (ie the more time or calories it took to reach the site or the more difficult the trip was), the less often a site was chosen.

Hedonic pricing

The hedonic pricing (HP) method, like the travel cost method, requires an assumption of weak complementarity[4] between a market good and the non-market good. However, HP operates through changes in observed prices rather than changes in quantities of the related market good(s) purchased. The market price of a good or service is known, and this price may be related to attributes of unpriced environmental goods or services that can be estimated. The HP approach is commonly used in housing markets where one may be interested in the values of the characteristics of houses or neighbourhoods in relation, for example, to air pollution, noise levels or aesthetically pleasing views. In this case, a house is viewed as a bundle of characteristics and purchasers reveal their preferences for these characteristics through their purchases of particular houses. Similarly, this approach could be used to disaggregate the market price of a safari hunting trip in order to estimate the value of an environmental attribute associated with the trip (eg types and quantities of animals killed/seen, habitats experienced during the trip, etc). The assumption underlying this approach is that values of these characteristics are capitalized in the trip or house prices. Thus, the basis of the values of trip or house characteristics is equivalent to the differential prices purchasers pay. HP involves the application of econometric techniques to data

from these types of private goods markets; by using the characteristics of the goods and prices, it is possible to derive estimates of the implicit prices of the characteristics.

Using the HP method involves a two-stage process. The first is the establishment of the hedonic price function. This involves the specification and estimation of a regression equation in which the dependent variable comprises the house or property prices and the independent variables are the characteristics of interest. The coefficient on each characteristic from the regression model assesses the responsiveness of the property price to a change in that characteristic (the partial differential or marginal price). A challenge exists in determining the relationship between price and a change in the characteristic. For example, the marginal prices are not necessarily constant or linear and could decline as the quantity of the characteristic supplied rises. The marginal price of one characteristic could also be related to the levels of other kinds of characteristics. All of these concerns relate to the functional form of the regression equation and the presence of collinearity among independent variables in the analysis. These are typically econometric issues and the best way to proceed is to examine a variety of functional forms for the hedonic price equation and to test which form works best.

Implementing the first step of the HP method is problematic in developing countries for a number of reasons. Firstly, the analyst needs a set of sale prices associated with actual market transactions. Secondly, the characteristics of the properties or goods in question (eg number of rooms), as well as the environmental characteristics of interest (eg water quality), need to be available and measured in appropriate units. This can be difficult in developing countries because knowledge about pollution levels may be low. Thirdly, the property market is assumed to be in equilibrium. This assumption arises, in turn, from the underlying premise that the environmental characteristics of properties are superior over individuals' own valuations of properties. In other words, cultural features of property ownership and characteristics may not be included in the hedonic model.

The second stage of the hedonic method is the identification of the bid function. The necessity of the second stage arises from the fact that the hedonic price function results from observations which consist of points of equilibrium between supply and demand (eg the property market is in equilibrium) (see Box 4.1 for a discussion of some of these concepts). However, since the interest on the part of the analyst is benefit estimates, which are derived from the demand (or bid) function, this function must be revealed. Only under some very restrictive conditions (ie every individual has identical tastes and preferences and the supply of the properties is fixed in the short run) will the hedonic price function represent the bid function. One cannot overstate the difficulty of identifying the bid function. The basic issue is that all individuals in the market face the same equilibrium price schedule (ie the hedonic price function), and an observed purchase by one of these individuals only reveals one point on their bid function.

One solution to this problem is to regress the implicit price estimates from the hedonic price function on those physical characteristics of properties and individuals' socio-economic characteristics which one thinks influence the demand for the properties being examined. Others have simply assumed that consumers have homogeneous tastes and preferences. Box 4.10 provides a developing country example where the researchers stratified

the market into three income groups and assumed that individuals within each stratum have identical preferences.

The hedonic pricing method is rarely used in developing countries. The research paper by North and Griffin (1993) described in Box 4.10 is one of only a few applications. The major reason for the lack of developing country applications lies in the lack of information on prices of properties or market goods. In some cases this scarcity of information may be the result of the lack of markets for some properties, or the use of barter and exchange in which information on transactions are difficult to uncover. However, the deficiency of information extends also to the characteristics of the goods or properties of interest as well. For example, it may be difficult to determine what the characteristics of houses were that were bought or sold, and even more difficult to uncover information relating to the environmental features of interest associated with the properties. However, there is ample room for further exploration of this method in developing countries. For instance, room rates in Zimbabwe for hotels of very similar quality vary by a factor of five to ten around the country. Much of this variation is probably related to the sites in the vicinity of the hotel, thus opening up a means for valuing such sites.

Related goods approaches

There are a number of valuation methods based on the premise that estimates for prices of certain goods may be derived from prices for substitutes for those goods (see also the section on 'Using own-reported values as the basis for valuation' in Chapter 2). These various replacement value methods or substitute approaches are somewhat crude and fairly data intensive; but they can provide policy-makers with useful estimates of the value of forest

Box 4.10 Using hedonic property valuation to value water source as a housing characteristic

North and Griffin (1993) used the hedonic pricing method to estimate the WTP of households living in a rural area of the Philippines for particular water-supply services. This research was important in that improving water-supply services in developing countries is frequently proposed and involves communal sources that are usually unpriced. The authors were interested in comparing household preferences for communal sources with alternative sources, such as private connections at residences and taps or wells on properties. For communal supply, the authors were interested in the value that households place on distance to the public or communal source of water.

Photograph by Chris Lovell

Figure 4.6 *A small dam in southern Zimbabwe relying on surface runoff, the silt content of which is dependent upon the management of the catchment. The estimation of the value of water and the value of the catchment-protection services provided for by woodlands and grazing areas could be tackled by using techniques presented in this chapter*

Data for the hedonic model came from a survey of approximately 1900 households chosen at random in the Bicol region where incomes are among the poorest in the Philippines. The sample was drawn such that the survey results were representative not only of the region, but also of the local distribution of income. About 90 per cent of those sampled actually owned their homes and about 2.5 per cent paid rent. A respondent, who was the head of the household, was asked to provide an estimate of the value of the actual dwelling (not the land surrounding the home) or the monthly rent, if appropriate. The survey included an extensive list of questions about the characteristics of the dwelling, such as number of bedrooms, location, current water source, etc. Additional questions involved information that was used in a human-capital formation model to estimate the permanent income of each household.

The authors developed the hedonic price model by estimating a 'bid-rent function'. This function involves the assumption that the purchase or payment of rent for a dwelling is based upon that dwelling's characteristics. This model involves regressing the dwelling's prices (or rents) on the various characteristics, including one of six types of water sources observed in the sample. The marginal WTP for each characteristic is the derivative of this estimated regression equation with respect to that characteristic. If a single equation were estimated on the set of prices and characteristics collected from the Bicol region, the authors would have to assume that all of the households in the region had similar preferences. Given that the sample was characterized by a wide income distribution, the authors addressed this assumption by dividing the sample into three different income groups (low, medium and high) and estimated a different bid-rent function for each group.

The resulting models suggested that household preferences were consistent with expected relationships. For example, households in all income groups were willing to pay more for dwellings constructed of good-quality materials; and households in middle- and upper-income groups, but not low-income groups, were willing to pay more for additional bedrooms. Regarding private water supplies, the results suggested that all households would pay about half of their monthly imputed rents to have water piped into the dwelling by a public-supply system. High-income households would be WTP US$1.95 per month over and above their existing monthly costs of water for this service. Middle-income households would be WTP US$2.25 per month and low-income households US$1.41 per month for this same service. Middle- and high-income households value a well or piped water in their yards, but not as highly as piped water in the dwelling. For public or communal water sources, only high-income households would be WTP a very small but positive amount for a water source closer to their residence.

This study shows that the households in this region of the Philippines place a value on water sources and that this value is capitalized into the price of their dwellings. However, the authors compared these values with the costs of actually providing improved water-supply services to the households, congruent with their preferences, and found that the costs were much higher. This information suggests that for this region, improving the quality and proximity of communal sources of water would be inappropriate.

resources for subsistence use. For some goods, the alternative to replacement-value methods is to have no known values for such goods, and from a policy perspective, it may be better to have rough estimates than no data at all.

Replacement-value or substitution methods are particularly effective in areas with mixed economies, where individuals can choose between self-provisioning or purchasing goods. A considerable number of these studies have been conducted in the Canadian north where Aboriginal people have access to purchased goods (particularly food), but also widely practice subsistence harvesting of food, medicine, materials for clothes and shelter (Tobias and Kay,

1993). There have also been numerous examples of this approach in developing countries (eg Campbell et al, 1991; Directorate of Forestry, 1996, Box 1.1; Shackleton et al, 1999a, Box 1.2)

The logic behind the related goods substitution method is simple. The underlying premise is that the welfare provided by a harvested good is closely related to the welfare provided by the substitute purchased good. There are some problems with this assumption that will be discussed later in this section. Some goods have both market and non-market uses. Other goods have domestic or subsistence uses but have market-based substitutes from which one can derive values for the subsistence goods. The first method in this suite of tools, the direct-substitute approach, is the most preferable. If a forest good, such as a fruit or fuelwood, is used for domestic purposes but has direct substitutes (eg the same type of fruit or fuelwood imported from neighbouring markets), then the value for domestically harvested and used goods may be imputed from the market price of the substitutes. All other things being equal, the closer the substitute, the greater the accuracy in the imputed value. In the fuelwood example, valuing self-provisioned firewood with purchased firewood of the same species makes for the best comparison. Alternatively, one could value self-provisioned wood with purchased firewood of a different species. One could calculate the BTU (British thermal units) differential between the species and make a more accurate valuation by taking such differences in wood quality into account. In some cases, the next most reasonable alternative will be kerosene, charcoal or heating oil (see Box 4.11).

The direct-substitute approach is commonly used with food. Harvest data are collected for wild game, fish, fruits, berries, roots and other plant material. Where data are collected in local units (eg baskets), conversion factors are determined through observation or by directly measuring the average weight and the usable portion of various foods. The manner in which harvested goods are accounted for depends upon the nature of the substitute good. For example, weights for red meat may be based on the dressed weight of a harvested animal if the purchased substitutes are commonly sold as individual cuts of meat. Conversely, the weight of the whole bird is the appropriate substitute if chickens purchased live at a market are substituted for wild harvested waterfowl. Once the appropriate substitutes are determined for game animals or plant material, the valuation is simply a matter of imputing the price of the

harvested good from an equivalent amount of the purchased good.

The barter-exchange approach is a variation of the direct-substitution method. The barter-exchange method is applicable when the good in question does not have a direct market-based substitute, but when a value for the good may be determined indirectly through its value as a bartered item. For example, suppose a given forest product, a particular sort of fruit, is consumed directly and bartered but not sold in any markets. One could take the average value of a commonly bartered item that is exchanged for the fruit and thus derive its value indirectly. Say one fruit is worth a third the value of one dozen eggs, and one dozen eggs are worth $1. The fruit then can be assumed to be worth $0.33. Other methods are related to these replacement methods, depending upon locally available substitutes, the degree to which the local economy is a mix between self-provisioned, bartered and purchased goods, and other factors. An example of a barter exchange game that can be conducted during participatory rural appraisal is given in Box 6.8.

There are a number of challenges and difficulties associated with using this suite of non-market valuation methods. One problem, from a logistical point of view, is simply documenting the amount of material consumed. Data may be gathered through direct observation or through surveys (see Chapter 2 for a detailed critique of survey-based approaches). When using the survey method, it is critical that researchers have detailed knowledge of the local subsistence economy, or have significant input from community members in the design of the survey instrument so that full range of goods harvested is accounted for. Both surveys and direct observation are time consuming, expensive and data intensive.

BOX 4.11 REPLACEMENT VALUES FOR FOREST RESOURCES IN A MIXED ECONOMY

Fort Liard and Nahanni Butte are isolated Aboriginal communities in the Northwest Territories of Canada and have a combined population of 570 individuals. A study by Beckley and Hirsch (1997) on bush use of natural resources in these communities concluded that between Cnd$900,000 and Cnd$1.7 million of value were derived from forest use in the harvest year of 1993 to 1994. The category of goods with the greatest value was meat harvested from the bush, which accounted for 54 per cent of the bush harvest. A variety of valuation tools were used. Some products (birch-bark crafts, tanned moose hides and other furs) were sold in markets, while others were estimated through replacement values. Other forest attributes could not be estimated at all, such as medicinal plants, building materials and the spiritual and cultural value of the forest.

Calculating replacement values for fuelwood and berries was straightforward. The climate is very cold and the only substitute available for fuelwood locally was heating oil (see Table 4.4). Based on a conversion rate of 414.1 litres per cord of wood, a value for fuelwood consumption for the population of both villages was estimated. The sample in Fort Liard consumed 693 cords and Nahanni Butte consumed 222 cords. Using the price per litre of heating oil (Cnd$0.59 in Nahanni Butte and Cnd$0.45 in Fort Liard), the value of fuelwood for the villages was Cnd$128,850 for Fort Liard and Cnd$54,238 for Nahanni Butte.

Table 4.4 *Calculation of fuelwood values using replacement values*

Product: fuelwood	Quantity (cords)	Substitute good (heating oil @ Cnd$/litre)	Value (Cnd$/cord)	Replacement (Cnd$)
Fort Liard	693	0.449	185.93	128,850
Nahanni Butte	222	0.59	244.32	54,238

Using a similar methodology for berries, a value for 590 litres of berries harvested in the two communities was calculated. There were two different prices in local stores for blueberries and raspberries. Most of the berries harvested were blueberries or species that had similar physical properties to blueberries (400 litres). For these, the blueberry price was used (see Table 4.5). Raspberries were more unique and more expensive. For these, the store price for raspberries was used. The store prices that were used were for frozen berries, and obviously the berries harvested in the bush were fresh when harvested. Most of the berries harvested were ultimately frozen, however, making the substitution more appropriate and, therefore, the valuation more accurate.

In the end, calculations were performed for the value of meat and fish harvested, crafts sold in the local shops, hides and furs sold, fuelwood and berries. In each community over half of all households were interviewed and the data were extrapolated to the entire population. Given considerable differences in household participation in the bush economy, the final results were presented as a value range rather than as a single number. In addition, the estimates were conservative given the inability to calculate values for a wide range of non-timber, non-market forest values, such as construction materials, medicinal plants, cultural and spiritual values and others. For some of these goods the constraints to recording values were methodological. For others, constraints were merely a matter of resources for the research. Other aspects of the study explored the per cent of the bush harvest that was shared with non-nuclear family, elders and friends, and the distributional impact of changes to the local economy as resource extraction increased in importance.

Table 4.5 *Calculation of fruit values using replacement values*

Product	Quantity (quart)	Substitute good	Value (Cnd$/quart)	Replacement (Cnd$)
Raspberries				
Fort Liard	172.52	Frozen raspberries	6.64	1145.53
Nahanni Butte	19.10	Frozen raspberries	6.64	126.61
Blueberries				
Fort Liard	363.20	Frozen blueberries	5.68	2062.97
Nahanni Butte	35.41	Frozen blueberries	5.68	201.14
Total				3536.25

Given varying degrees of participation in market and subsistence economies among members of the same village or community, care must be taken if a sampling approach rather than a full census is taken. If one interviews only the active harvesters and extrapolates values from that sample to the larger community, one may overestimate the value of the subsistence harvest.

Another complication in replacement value methods stems from the difficulty in determining 'equivalent amounts' of purchased or bartered goods. In some cases it may be appropriate to consider nutritional equivalents (calories or protein – see Box 4.9), while in other instances it may be appropriate to consider weight or volume or BTUs. Furthermore, one can imagine the complexity of using this method in situations where spatial distribution of resources (eg fuelwood) plays a large role in influencing value. In this case, caloric expenditures will differ depending upon the location of the fuelwood and the fuelwood collectors. Accordingly, travel cost methods (see Box 4.9) may be more appropriate for valuing these types of resources.

A third complication may also arise with this method and involves goods that have multiple uses and provide multiple products. For example, the flesh of a mammal may be used for meat, the hide may be used for clothing and various parts may be sold in legal or illegal markets for cash. All of these various 'accounts' must be tracked and summed. Similarly, with plants, a single harvested tree may provide forage for livestock, wood for tool or craft-making and fuelwood.

Many traditional cultures express a preference for wild game over purchased substitutes. This preference is, in part, due to taste differences, but also due to the importance of tradition and the connection to traditional culture that flows from harvesting and preparing traditional foods. Hunters and gatherers achieve status through their self-provisioning exploits that they are not likely to achieve from earning a wage or purchasing meat or vegetables from a store or market. This extra value attributable to harvested goods is not accounted for in replacement value studies.

For some goods harvested it is very difficult to determine appropriate substitutes. This is particularly true of medicine and traditional remedies. In many cases there are no locally available substitutes. In other cases, it is difficult to attribute the benefits received or welfare gained from chewing a root to relieve a headache or cramps, when compared to taking a pill.

Despite all of these challenges and complicating factors, replacement value methods may still provide significant estimates of the value of local subsistence activities. Often, policy-makers assume the value of activities to be zero. Converting these goods into currency equivalents helps make the point that there are opportunity costs associated with developments or policies that compromise a community's or village's ability to provide for some of their own nutritional, fuelwood, shelter and clothing needs through self-provisioning.

Conclusions

People make hundreds of choices every day, with varying degrees of information. Societies also make choices. In order for society, and the political decision-makers who make choices on behalf of society, to make the best choices they need information. Non-market valuation methods can provide some of that information. In the past, the economic information that influenced a decision on whether to build or not to build a dam was limited to the value of the standing timber in the proposed flooded area, and perhaps the costs of relocating people who lived in the area. Simply put, decision-making with this level of information will ultimately undervalue the forest and could result in a poor decision being made. This is particularly true if local residents derive subsistence from the forest; if people from other regions come to visit that forest; if people on the other side of the globe value rare or endangered species in that forest; or if local and non-local people simply value the existence of that forest. Non-market valuation tools, however imperfect and incomplete, attempt to create a total economic value for a given resource. In other words, they attempt to provide decision-makers with much more information than is available from monetary markets.

In the developing world, the livelihoods of numerous people are directly tied to natural resources. Nevertheless, the value of those resources to the individuals and households who depend upon them is rarely quantified. The implications of this are significant. Decisions about the allocation of natural resources will go on, with or without quality information about the value of those resources to local people. In the policy world, there is an adage that 'you can't manage what you can't measure'. Non-market valuation methods represent strategies for measuring values that were not considered, before, in policy decision-making.

This area of social science is still developing. Some of the methods are considered controversial, and some analysts are concerned about the precision of the available methods. There are significant challenges in translating developed world models and methods to developing world contexts. Many strategies that work in the developed world (eg mail and phone surveys) are unworkable in developing world contexts. However, some researchers have enjoyed 100 per cent response rates in delivering non-market value surveys in the developing world. Different labour costs, literacy levels and access to research sites, for example, have and will continue to effect the way non-market valuation studies are carried out in the developing world. Clearly, there is value in attempting to elicit and quantify the broad range of benefits that natural resources supply to local and non-local users. The more information we have regarding the value of forests and trees, the better decisions we will ultimately be able to make regarding their allocation, conservation and use.

Notes

1   Note that this payment can be positive or negative. The sign of the WTP amount reflects whether the consumer gains or loses from the change. If this amount is negative then the payment is commonly referred to as the willingness to accept compensation (WTAC).
2   Another component of total economic value mentioned by many others is the value of preserving options. Some consider this to be part of use value (eg Georgiou et al, 1997). However, recent theoretical work recognizes that the value of preserving options arises from uncertainty about future demands and/or future supply (Freeman, 1993). In general, the literature on the preservation of options is complex and is beyond the scope of this chapter.
3   This term refers to a relationship between the purchase of a market good(s) that must occur in order to enjoy the benefits of a quasi-public or public good. For example, in order to visit a national park, one must purchase petrol and other items to visit it.
4   See Note 3.

Chapter 5

Economic decision-making frameworks for considering resource values: procedures, perils and promise

Terrence S Veeman and Martin K Luckert

Introduction

Despite somewhat different perspectives, rooted in their respective biophysical and social sciences, biologists and economists have much common cause in wishing to see natural and environmental resources managed and used for the welfare of present and future generations. As other chapters in this book show, an understanding of resource values is crucial to our understanding of livelihoods and the subsequent planning of projects and policies that attempt to enhance rural livelihoods. The purpose of this chapter is to present some alternative ways that these values can be used within decision-making frameworks in order to provide an understanding of rural livelihood systems and provide policy-relevant information.

The following section presents three different ways that resource-value information can be used. Although these three frameworks may all be considered variants of cost-benefit analysis, the differing uses of cost and benefit information warrant classifying frameworks into three areas. These areas differ primarily according to the specific group of people whose welfare is being evaluated. Following the presentation of the three frameworks, the section on 'The background essentials of using benefit and cost information' reviews some basic concepts common to all three types of assessments. This necessarily introduces the concept of discount rate and describes different kinds of decision criteria (eg net present value, benefit-cost ratio). As part of the discussion of these basic concepts, the section highlights how these ideas may vary depending upon the framework being used. The final section cautions about the uses of these frameworks.

Three frameworks for considering benefits and costs

Variants of cost-benefit analysis may be categorized into three general frameworks: development-project assessments; 'green'- accounting assessments and household assessments. The following sections summarize each of these.

'Green'-accounting assessments

There has been increasing recognition that the state of natural resource endowments over time is crucial to sustainable livelihoods. Therefore, benefit and cost information may also be used in green accounting exercises to assess whether these endowments are increasing, declining or being maintained. Assessments may be quite localized, assessing resource endowments for a single village, or may be undertaken to supply information for national accounts (see the sub-section on 'Changes in the stocks of renewable resources' in Chapter 2).

Historically, national accounting has been targeted towards tracking measures of economic activity, such as gross national product (GNP) or gross domestic product (GDP), over time. However, such measures are deficient on at least two major counts. Firstly, as concerns have increased regarding the environment, more attention has been paid to the role that natural resource endowments play in fuelling economic activity and in providing environmental services. Accordingly, current trends are to supplement traditional economic-activity measures with information regarding how resource endowments change over time and how environmental quality may be changing.[1] The key adjustment involves the inclusion of the depreciation of natural capital. If resource endowments are declining, it is important to weigh these costs against benefits arising from increases in economic activity, and to assess what declining endowments may mean for future economic growth. Green-accounting adjustments have been made for the economies of Mexico and Papua New Guinea by the World Bank (Serageldin and Steer, 1995). In Mexico in 1985, for example, green or environmentally adjusted net domestic product was estimated to be 87 per cent of the conventional national domestic product (NDP) figure. Secondly, measures of economic activity have not traditionally included the often substantial benefits of non-market or informal sector activity that economies may generate. For example, the 'hidden harvest', consumed in subsistence economies, is not picked up in national accounts. Similarly, non-marketed benefits, such as the value of water in rural areas, are also omitted (see the section on 'When own-reported values cannot be used: assumptions, omissions and other techniques' in Chapter 2).

Green-accounting assessments differ substantively from development-project assessments. Whereas resource-development assessments concentrate on assessing the welfare implications of specific potential projects at a given point in time, green-accounting assessments track changes in economies over time that may have numerous or no development initiatives. Accordingly, green-accounting frameworks are generally set up with replication in mind, as annual assessments are generally anticipated.

Development-project assessments

This framework refers to a situation where a government or donor agency is considering a major project, such as building a dam, or converting woodlands to agriculture. The question that is considered is whether the investment under question is advisable. Such considerations could be in the context of debating alternative land-use polices – for example, with respect to land-use policies in Amazonia (see Box 5.1). The basic purpose of such assessments is to provide information about welfare changes to the people affected by the project. This information can be used for three basic purposes.

BOX 5.1 COMPARISON OF THREE LAND-USE STRATEGIES IN AMAZONIA

This box compares the economics of three types of land use currently practised in the Brazilian Amazon: forest extraction, extensive agroforestry and intensive agroforestry (Anderson, 1992). The data from the forest extraction case comes from an extractive reserve in which 67 rubber-tapping families live, with an average holding size of 372ha each (see Table 5.1). Households extract rubber and Brazil nuts from the forest, and are remote from markets. Factor inputs are extremely low, averaging 0.53 person-days of labour and US$0.24 in materials per hectare per year. Economic returns have been calculated in alternative ways; these reflect the total income derived from the enterprises in relation to the various inputs. The returns for the forest extraction case are low. The average net annual return is US$872 per holding or US$2.35 per hectare. Expressed in terms of family labour (US$4.38 per day), the returns are greater than the prevailing wage rates (US$2.60). The net return per unit capital invested is quite high (US$10.75).*

Table 5.1 *Economic comparison of three land-use strategies in the Brazilian Amazon* *

	Forest extraction	Extensive agroforestry	Intensive agroforestry
Average holding (ha)	372	36	28
Factor inputs			
of labour (days per ha per year)	0.53	18.36	88.46
of capital (US$ per ha per year)	0.24	5.16	720.29
of capital (US$ per holding per year)	89	186	20167
Returns			
US$ per holding per year	872	2733	9499
US$ per ha per year	2.35	75.93	339.27
US$ per day of labour	4.4	4.18	3.84
US$ per unit of capital	9.79	14.72	0.47

* By using annual values, net present values are not needed. However, there is an implicit assumption that intertemporal flows of benefits and costs are non-variant.

In the extensive agroforestry case, holdings are only, on average, 36ha. A typical holding is surrounded by a house garden, a zone of managed forest and a zone of unmanaged forest. This land-use system is characterized by greater investments and returns than the forest-extraction land use. Much greater labour inputs are required per hectare, and capital investment per holding is double that for forest extraction. Net returns per holding and per hectare are much higher, and are slightly higher per unit of capital, while per unit of labour they are about the same.

The third case, intensive agroforestry, is characterized by extraordinarily high inputs of labour and materials. The native forest cover is replaced by plantations of tree crops in complex spatial arrangements. Farmers market a wide range of products to a nearby market. The net returns per holding and per hectare are considerably higher than the other land-use options; but net return per unit of capital is considerably less.

The principal advantage of extraction is its low requirement for capital. However, this also results in low economic returns, making this land-use option highly susceptible to substitution by more intensive land uses. The most intensive system requires high factor inputs and ready access to markets.

* The need to be explicit in describing methods has been made in the Introduction to Chapter 2. There are many cases where it is difficult to draw conclusions from data or to use the data for comparative purposes because of lack of detail in the methodology. In this example it was difficult to assess whether the returns were calculated on the basis of total income or net income (eg net of capital and/or labour costs). Either way, expressing returns in terms of one physical input can be confusing, as it could be interpreted that the income is a return to the one input, when it is actually a return to all inputs.

Firstly, development-project assessments can inform the decision on whether a project should go ahead. Although there are numbers of available welfare measures and criteria for making decisions about whether a project improves welfare (see the section below), the general approach is to call a project 'feasible' if the benefits of the project exceed the costs. This has been the primary use of development-project assessment.

The second purpose of development-project assessments involves prioritizing. In some situations, resource-development assessments may disclose that there are multiple projects, which all all appear feasible. Frequently, however, the funds available are not sufficient to undertake all feasible projects. In such cases, cost-benefit information can also be used to rank the set of feasible projects.

Finally, development-project assessments can be used to investigate variants on proposed developments. Such analyses can provide information on how changes to the project may alter its potential contribution. For example, the cost-benefit analysis of an afforestation programme in Zimbabwe never took into account the prior question of whether increased wood production could be obtained more cheaply and equitably from promoting the management of indigenous woodlands. An analysis of different approaches to grazing is illustrated in Box 5.2.

Sometimes, despite the search for feasible development projects, the results of the analysis may indicate that no development should occur. This notion dies hard with some foresters who see trees in a forest, or some engineers who see water running down a river, as equivalent to 'wasted' physical resources. However, if the cost of the labour, capital and other resources needed to exploit the resource exceed the benefits that the resource provides, non-use of the resource may be the wisest course of action.

Household assessments

The third framework is used on a much more micro-scale. These assessments involve investigating the benefits and costs, received and borne, by individuals or households.[2] For example, Box 5.3 investigates household returns from planting a medicinal plant in Zimbabwe, while Box 5.4 investigates household returns to beer and basket production in Botswana. The reason behind such micro-assessments is generally to gain a better understanding of household livelihoods on a local scale, as opposed to higher aggregations of welfare estimates found in the first two frameworks. While development-project and green-accounting assessments consider changes in costs and benefits within an identified region, there is no information provided about how these changes in benefits and costs are borne by households.

Household-level information regarding benefits and costs can be used for several purposes. Firstly, such information is valuable from a behavioural perspective. An accounting of benefits and costs at a micro-level can provide insights into why individuals or households adopt, or fail to adopt, potential development activities. For example, in Box 5.3 a cost-benefit analysis is used to show why it is likely that households in south-eastern Zimbabwe will plant *Warburgia salutaris*, a valuable medicinal plant. Secondly, household results can be scaled up to estimate regional effects. For example, average benefits and costs over a number of households may be calculated, and then scaled up according to a sample of households to derive information about a region under scrutiny (see Box 5.5). Finally, household-level data provide insights into

Box 5.2 Using net present value to compare different livestock strategies

Bruce Campbell and Martin Luckert

One of the major service functions of the savannas and woodlands of Africa is that of a provider of browse and graze for cattle, as well as other livestock. During the last decade a 'new rangeland science' has emerged (eg Behnke et al, 1993). One of the tenets of the new science is that pastoralists should not adhere to a single conservative stocking rate, but rather adopt an opportunistic strategy, where numbers will fluctuate widely in response to good and bad seasons. It is further argued that opportunistic strategies give higher economic returns compared to strategies based on conservative stocking rates. Campbell et al (2000a) have compared the benefits and costs of four cattle-management scenarios. The analysis is based on a simulation model of the fluctuation, over time, of animal numbers, outputs and prices. The decision criteria used is that of net present value (NPV), which is appropriate given the fluctuating nature of the stream of benefits over time (driven largely by rainfall patterns and the subsequent collapse and recovery of animal numbers).

The results suggest that strategies based on conservative stocking rates will have higher NPVs than strategies based on opportunistic stocking rates (see Table 5.2). However, to receive the full benefits of destocking, a decision to destock has to be made at the level of the community, because the benefits of improved outputs can only be achieved if the stocking rates of the communal grazing lands are reduced. Making collective decisions about managing numbers is a process with considerable transaction costs (Ostrom, 1990). It is surprising that a tight tracking scenario (where numbers of cattle are managed by purchasing and selling in order to maintain numbers in equilibrium with the available feed resources) is recommended in the relevant literature (Behnke and Kerven 1994). The results suggest that such a system will come with considerable economic losses.

Table 5.2 *Comparisons among alternative livestock scenarios in terms of their overall net present value (Z$million)*

Assumptions Discount rates	Opportunistic scenario	Conservative scenario
Chivi		
8 %	20	47
17 %	(7)	58
25 %	(94)	53
Mangwende		
8 %	40	54
17 %	53	86
25 %	54	127

Note: All values are net present values for a 15-year stream of costs and benefits in Z$million. Brackets = negative values.

Only one of the simulation runs is presented here. A large number of assumptions were varied in order to determine the sensitivity of the results.

Box 5.3 *Muranga* returns: the economics of producing a rare medicinal species reintroduced in south-eastern Zimbabwe

Terry Veeman

The bark of *Warburgia salutaris*, locally known as *muranga*, is a very important traditional medicine in southern Africa. However, trees of this species are at, or near, extinction in the wild in Zimbabwe. In this box, the economics of reintroducing this rare medicinal species is examined (Veeman, T et al, 2001). To date, only very limited plantings have occurred (less than 100), essentially as small pilot projects involving a few traditional healers/growers and some small farmholders in a relatively moist, hilly and remote ecozone in south-eastern Zimbabwe.

In order to investigate costs and benefits associated with *Warburgia*, a number of assumptions regarding its management were made. Firstly, it was assumed that the length of the planning period would be 24 years. Secondly, the harvesting regime was viewed to be a cut-and-coppicing regime, rather than one involving the selective removal of patches of bark or branch bark from the maturing tree. In the chosen baseline scenario, it was assumed that the growing tree would be harvested at year 8, yielding 6 kilograms of air-dried bark; thereafter, coppices would arise and be cut every fourth year, generating a bark yield of 4 kilograms in each of years 12, 16, 20 and 24. The underlying rationale for such bark-yield assumptions was the measurement of a small (n = 6) sample of known-age *Warburgia* trees and the analogy to wattle, a species for which specific bark-yield information is available. Although *Warburgia* is a shorter tree than wattle and grows more slowly, it is regarded as having thicker bark relative to wattle (Cunningham, pers comm, 1999). In addition, to cover the possibility of overoptimistic bark-yield assumptions, scenarios were also modelled using one half and one quarter of baseline bark yield.

The assumed price scenarios were based on empirical evidence collected in May 1999 at three Zimbabwe markets: Harare, Mutare and Chipinge area, where 16 *muranga* bark samples were purchased with a mean average value of Z$1150 per kilogram. This equates to approximately US$33 per kilogram, using an exchange rate of Z$35 to US$1. It was assumed that the relevant farm-gate price of *muranga* bark would be some 70 to 75 per cent of the foregoing retail level values, given that the bark is readily transported and stored. Furthermore, it was assumed that *muranga* price levels would remain the same, in constant dollar terms, over the planning horizon. Consequently, the three *muranga* price levels assumed in the analysis were as follows:

1   low or pessimistic: US$5 per kilogram;
2   medium or baseline: US$25 per kilogram; and
3   high or optimistic: US$50 per kilogram.

The low price scenario was designed, in part, to reflect the possibility that marketing margins are much higher and farm-gate prices are lower than assumed in the baseline model. As well, this was to accommodate the possibility that large-scale commercial production might arise, adversely affecting price levels for the remote smallholders with restricted plantings.

Returns and costs were generated on a per tree basis for both traditional healers and small-scale farmers. In addition to bark values, it was also assumed that healers and, to a lesser extent, farm households, would gain modest benefits from the use and/or sale of leaves. After receiving very small initial benefits, healers were assumed to receive US$0.66 per month per tree from the use of leaves, while farm households would receive an imputed gain of US$0.23 per month per tree (beginning in year 4).

Inputs and harvesting costs, primarily labour, were also included, utilizing an agricultural wage rate of Z$2.57 (US$0.07) per hour. An important initial cost is the purchase of seedlings. The cost of seedlings was subsidized considerably in the initial pilot planting, costing the grower only Z$10 per seedling (a result of a WWF People and Plants Initiative). In this social analysis, it was assumed that the cost of a seedling, which could be relevant to future planting, would be five times higher at Z$50.

Finally, benefits, returns and costs were discounted using a number of real discount rates. For the small-scale farmholders, who faced considerable risks, the individual or private rate of time preference (discount) was assumed to be 17 per cent, a rate in line with other farm-level studies in rural Zimbabwe. For the traditional healer, typically a higher income household with more assets and less risk, the private discount rate chosen was 10 per cent. In the social cost-benefit analysis, the chosen social rate of discount was 5 per cent, reflecting social time-preference arguments related to long-lived investments such as trees.

The empirical results of the cost-benefit studies for several scenarios for private discount rates are reported in Table 5.3. In each scenario, a summary benefit-cost ratio (in private terms) is shown for both the case of the traditional healer (*n'anga*) and the case of the smallholder farmer. The results are extremely robust, evidencing very high benefit-cost ratios at both the healer and farmer levels, particularly under the baseline price and base-yield scenario, and even more so under the optimistic price and base-yield scenarios. For example, under the baseline price and yield assumptions, the private benefit-cost ratio relating to the growing and harvesting of *Warburgia* tree is over 42 to 1 for the healer and over 24 to 1 for the smallholder. Even with the low price and one-quarter base-yield scenario, the benefit-cost ratios remain relatively high at 13.3 to 1 and 4.5 to 1 for the healer and farmer, respectively.

Table 5.3 *Benefit-cost ratios for alternative scenarios regarding* muranga *bark price and yield (per tree basis), Mount Selinda area, Chipinge District, south-eastern Zimbabwe (private discount rates are used in these scenarios)*

	Price US$/kg	Yield of bark per tree (main crop, coppice)		
		Base yield (6, 4) kg full-leaf benefits	One-half base yield (3, 2) kg full-leaf benefits	One-quarter base yield (1.5, 1) kg full-leaf benefits
Healer	5	17.9	14.8	13.3
Farmer		7.6	5.5	4.5
Healer	25	42.2	27.0	19.4
Farmer		24.2	13.8	8.6
Healer	50	72.6	42.2	27.0
Farmer		45.0	24.2	13.8

Muranga production is not only highly profitable in private terms but also in social terms. In the simplified social analysis, a much lower social discount rate (5 per cent) is used and the influence of seedling subsidies is removed. Even so, under the low-price and low-yield scenario, the social benefit-cost ratios remain relatively high: 10.3 at the healer level and 3.2 at the farmer level. Under stronger price and yield assumptions, the economics of *muranga* production is even more attractive both socially and privately. It is only under a worse case scenario – low bark yield (one-quarter baseline), low bark price (US$5) and no benefits from leaves whatsoever – that *muranga* production becomes socially unfeasible, and then only at the farmer level (BC ratio = 0.74).

This analysis strongly suggests that expanded *muranga* production, particularly in the context of smallholder agriculture in the remote, hilly region of south-east Zimbabwe, is clearly very economically attractive and conducive to improving rural incomes and livelihoods in that locale. It is certainly possible to 'kill the goose that lays the golden egg'; through larger-scale commercial production by commercial farmers, greatly expanded supplies of *muranga* could flood the market, depress the price and harm the smallholders. At the moment, however, there is considerable scope for the managed release of more *Warburgia* seedlings to the small farmholders in the study area.

who receives the benefits and who bears the costs of alternative activities. Important differences in welfare effects within and between households may be picked up through such analyses. For example, Box 5.4 shows the potential importance of basket-making to women, while the section on 'Emerging results' in Chapter 2 shows important differences between income categories of households.

Photograph by Mark Edwards

Figure 5.1 *Medicinal plant market in Belem, Brazil, with a wide variety of plant and animal products on display. Medicinal plants can often fetch high prices in urban centres, making it worthwhile to cultivate the plants, as illustrated for* Warburgia salutaris *in Zimbabwe (see Box 5.3)*

Box 5.4 Beer and baskets: the economics of women's livelihoods in Ngamiland, Botswana

Small-scale craft industries are often seen as a way of increasing the incomes of marginalized people, particularly women. Making use of available natural resources and extensive local skills in craft production potentially offers significant increases in local cash-income levels. In southern Africa, basketry, pottery, weaving or carving have developed new markets, often related to the tourist industry. This box examines the economics of basket production in two sites on the western edge of the Okavango Delta in Ngamiland, Botswana (Bishop and Scoones, 1994). Key activities within women's livelihoods were identified through a variety of participatory rural appraisal (PRA) exercises, including mapping (social and resource), matrix ranking, seasonal calendar diagramming, joining harvesting and collection activities in the field, and semi-structured interviewing. Key informants supplied the information required for the economic analysis.

Estimates of returns to labour in basket-making may be compared to the economic benefits that women obtain in other activities, in order to illustrate the relative importance of basket-making in their livelihoods and the likelihood of a good response to additional external support for the basket industry. Therefore, a range of estimates of returns to labour was collected for four major forms of employment: basket-making, agriculture, beer-brewing and drought relief. In the study area, basket-making is based on the *Mokola* palm leaf (*Hyphaene petersiana*), dye material (often the bark of the tree *Berchemia discolor*) and, in some cases, grasses or vines for making the cores of coils. Labour inputs are described in Table 5.4.

Table 5.4 *Average labour inputs to basket-making*

Production stage	Average hours per basket	
	Etsha	Danenga
Collection	2.8	2.0
Processing	0.6	0.8
Weaving	20.0	20.0
Total	23.4	22.8

The main economic activity of women (and men) in western Ngamiland is agriculture. Agricultural practices and crop choice vary mainly by proximity to inundated areas. Thus farmers living further from the flood plain (eg Etsha) primarily grow pearl millet on sandy soils and a number of farmers till by hand. Farmers living in or nearer to the flood plain (eg Danenga) mainly grow maize on heavy, alluvial soils and rely heavily on animal traction for tillage.

Beer-brewing is a common income-generating activity during the post-harvest season. A popular fermented drink is *Kgadi* beer, which many local women make from the fruit of the *Grewia bicolor* tree. The beer is sold at home, by the mug, to casual visitors. Estimated returns to labour in brewing *Kgadi* beer vary widely but average about 1.2 pula per hour, assuming an average of 7.5 hours to prepare and sell one batch. In recent years, both women and men have benefited from publicly funded 'drought-relief' programmes that offer casual employment for road-building and other local infrastructure projects.

To compare these different activities to basket-making, and to account for differences in the total amount of time spent on each activity, estimated daily returns are converted into hourly equivalents (see Table 5.5).

Table 5.5 *Returns to labour: baskets, beer, agriculture and drought relief*

Activity	Returns to labour: (income from sales-capital costs)/labour inputs (pula per hour)
Basket-making:	
Etsha site	0.33
Danenga site	0.42
Kgadi beer-brewing	1.21
Agriculture	1.19
Drought relief	0.75

Returns to labour vary widely among different economic activities. This is not altogether surprising. Since local labour markets are poorly developed and opportunities for paid employment are extremely limited, one would expect returns to labour to vary, even for a single class of labour. Nevertheless, it is striking that average returns to basket-making are lower than that of any other activity. A variety of interpretations are possible: the model of basket-making may have underestimated actual returns to labour (by overestimating the average weaving time, for example), or basket-making is a marginal activity relative to other income-generating opportunities. With respect to the second interpretation, a possible explanation for the relatively low returns is that basket-making is a low-risk activity that requires little in the way of capital investment, compared to the main alternatives. This hypothesis would be reinforced if it were observed that weavers are, on average, poorer than non-weavers. However preliminary observations suggest that this is not the case as a wide range of women in both villages were engaged in regular weaving.

Another possible explanation for the disparity in labour returns between different livelihood options is that they are not alternatives. Women do combine a variety of different activities. Basket-making and beer-brewing may not be directly competitive, because baskets can be woven while beer is sold. In the dry season, the major period when basket-making and brewing occurs, these activities are not competitive with agriculture. The interannual variation of options is also not taken into account in the aggregated figures presented in Table 5.5. Returns to agriculture drop dramatically in dry years, making alternative climate-independent activities, such as basket-making, more attractive alternatives.

BOX 5.5 CATTLE OR WILDLIFE?

Ivan Bond

During 1990 and 1991, a survey into the viability of cattle and wildlife enterprises in the large-scale commercial farming sector in Zimbabwe was undertaken from private and social perspectives (Jansen et al, 1992). The objective of the survey was to compare the viability of cattle and wildlife enterprises. These enterprises are located in semi-arid and arid regions of the country. In total, 139 ranchers were interviewed using a questionnaire survey.

Viability from a private perspective

The average financial return on investment showed that the profitability of wildlife enterprises increased from the semi-arid to the arid regions. Only in arid regions, however, did the average financial return on investment (internal rate of return) across all enterprises exceed 10 per cent, the level considered by the survey as profitable. The reasons for increasing profitability of wildlife production systems with increasing aridity were not clear and require further investigation. One suggestion is that wildlife production benefits from economies of scale, and therefore it is more viable on the larger ranches in the arid areas. In contrast, the average financial return on investment in all cattle was below 5 per cent in all regions.

The results of the survey have been substantiated by recent land-use changes on large-scale commercial ranches since the survey. Increasing numbers of ranchers are either starting to utilize their wildlife resources or are destocking cattle in favour of increased wildlife enterprises. The development of wildlife conservancies, in which economies of scale are realized through cooperative land and wildlife management between individual landholders, is tangible and substantial evidence of these changes.

Viability from a social perspective

For the economic analysis, two key variables – the price of beef and the exchange rate – were shadow priced (see the section on 'Problems in estimating benefits'). At the time of the survey the producer price of beef was controlled by the government and the costs of all tradeables were affected by the overvalued Zimbabwean dollar. Based on the net realization of the Cold Storage Commission's local and export sales, the producer price of beef was increased by 25 per cent, while the Zimbabwean dollar was assumed to be overvalued by 50 per cent based on parallel market rates.

The economic analysis showed that the government economic policy at the time negatively affected the viability of both wildlife and cattle producers. Using the economic return on investment, the analysis predicted that, in the absence of government intervention, both wildlife and cattle would be profitable land uses in all three natural regions.

The key economic indicator used was the domestic-resource cost ratio (DRC). The DRC is the ratio of the domestic factor costs in economic prices (G) divided by the value added (E–F) (revenue minus tradeable costs) in economic prices ($G/(E$–$F)$). An enterprise with a domestic-resource cost ratio of between zero and one is considered efficient, and one in which the country has a comparative advantage. The strength of the DRC is that it allows comparisons to be made between activities on the assumption that there was no government intervention in the market.

The ranch survey showed that 38 per cent of the 120 cattle enterprises, compared with 50 per cent of the 64 wildlife enterprises, were economically efficient (see Table 5.6). A further 34 per cent of cattle and 30 per cent of wildlife enterprises were nearly efficient. The results implied that, with the liberalization of the economy (ie the absence of government intervention in markets) the country had a stronger comparative advantage in wildlife production than cattle production. The implication of these results was that wildlife was a financial viable and economically efficient land use that, in a rational political environment, should be supported by government.

Table 5.6 *The distribution of the domestic resource cost (DRC) for cattle and wildlife enterprises*

	Cattle		Wildlife	
Total sample size	120	(100%)	64	(100%)
DRC>5	17	(14%)	5	(8%)
DRC 2-5	17	(14%)	5	(12%)
DRC 1-2	41	(34%)	19	(30%)
0<DRC<1	45	(38%)	32	(50%)

Discussion

The survey showed that there were both financially viable and non-viable cattle and wildlife enterprises, implying that there were factors such as ranch size, resource base, management skills and location that affected viability. It was, therefore, incumbent upon the ranch managers/owners to determine the optimum allocation of resources between cattle and wildlife production systems. Significantly, the diversification into wildlife was not an option available to all large-scale commercial ranchers unless there were sufficient 'core' breeding populations.

The influence of the survey and the results on policy towards establishing wildlife as a legitimate land use on large-scale commercial ranches have been, at best, uncertain. A political analysis of land-use systems conducted soon after the survey showed that wildlife was still not considered a politically legitimate land-use, making wildlife producers more susceptible to compulsory land acquisition under the resettlement programme (Hill, 1993). Latterly, analyses of the large-scale farming sector have recommended that land-use permits be introduced to control (and restrict) wildlife production (Hungwe and Fernando, 1996). In addition, the state has attempted to recentralize control of wildlife, thus jeopardizing the fundamental principles of devolution (Wild Life Act 1975) that originally facilitated the development of wildlife as a land use.

The background essentials of using benefit and cost information

Choice of accounting stance

Scale of analysis

The first major step in conducting cost-benefit analyses is the choice of an appropriate accounting stance (ie the scale of analysis). Is the analysis to be undertaken from a national, provincial, regional or household point of view? This choice is integral to the type of assessment framework that is used.

Typically, a provincial or national accounting stance is appropriate when provincial or national government departments are charged with managing natural resources, and when provincial or national monies are being used to fund specific projects. This has certainly been the case with a number of afforestation projects in a variety of African states.

When considering large projects, a major implication of choosing the accounting stance is that there may be beneficial impacts on certain areas or communities within the province or nation that would be included in a regional or local stance, but not in a provincial or national stance. In a national stance, many of the secondary benefits derived from project spin-offs cancel out, as local gains occur from diverting resources from other parts of the larger economy. In other words, the inclusion or exclusion of secondary benefits – economic activity that stems from, or is induced by, the primary activity – is intimately related to the choice of a suitable accounting stance. Once an accounting stance is chosen, the cost-benefit analyst compares:

- the state of the economy *with* the project; with

- the state of the economy *without* the project.

The analyst then makes sure that any benefit sphere does, in fact, represent a net gain in economic activity and not merely a transfer of economic activity within the provincial or national jurisdiction.

Accounting stances for development project or green-accounting assessments are frequently done at provincial or national levels. However, more micro-applications are also common, as are presented in this book. Despite the fact that most effort on green-accounting assessments has been at the national level, micro-level assessments of resource accounts could provide useful information regarding the sustainability of local livelihoods. Ecological information regarding natural resource stocks and flows may be easier to assess at these micro-levels, making accounting exercises more realistic biologically. In contrast to the above two methods, the accounting stance for household assessments is at a much more micro-level considering the benefits and costs to the households within a specified locale.

Private versus social accounting frameworks

Once the problem is bounded in terms of the reference group of people whose welfare is being considered, there are frequently at least two additional accounting stances specified: a private accounting stance and a social accounting stance.[3] In some development projects, it is expected that private companies will undertake the project. In these cases it may be important to know whether these companies will

153

receive sufficient, or more than sufficient, returns to their efforts. The benefits and costs that the private company will receive are disclosed from a 'private accounting stance'. This assessment uses returns and costs accruing from market transactions to see what the bottom line will look like for the participating company. This stance can also be taken for analyses based on households, as illustrated in Box 5.3 (which investigates the private incentives of households to plant a medicinal species).

In a cost-benefit study undertaken from society's point of view, other aspects should also be considered. Firstly, a social cost-benefit analysis must include external or unappropriated benefits (benefits which the private sector cannot capture) and external or uncompensated costs (costs, such as downstream siltation or environmental damage), which are imposed by the project or policy on third parties. Secondly, the market prices of outputs and inputs that are distorted should be corrected to reflect true societal scarcity values or opportunity costs.[4] Thirdly, in the case of potential large developments, social cost-benefit analysis should consider and estimate the economic activity generated by the primary activity that is linked 'forwardly' (as with processing and distributing industries) and linked 'backwardly' (as with input-supply industries). These stemming and induced secondary benefits are clearly important to regional income distribution and perhaps, also, under special circumstances, to welfare estimates on a wider scale (that is, to efficiency concerns). Finally, there is the distinct likelihood that social rates of discount will be lower than private rates (see the following section on 'Choosing a discount rate').

Like development-assessment projects, there may also be more than one accounting stance taken within household

assessments. Private accounting exercises may show whether it is in a household's interests to pursue a given activity, whereas social accounting provides information on whether or not it is socially desirable for a household to pursue an activity (see Box 5.3). Where an activity is socially desirable, but not adopted by the household, subsidies may be called for. Conversely, where activities are undertaken that are not socially desirable, taxes may be in order.[5]

Types of benefits and costs

In the enumeration and evaluation of benefits and costs, it is analytically useful to distinguish four categories or types of benefits and costs (Howe, 1971, p15):

1   Benefits and costs for which market prices exist and for which these prices correctly reflect societal scarcity values (for example, farm outputs that are not price supported and forest products, such as wild fruits, fuelwood and poles, where there are frequent transactions).

2   Benefits and costs for which market prices exist but for which the prices fail to reflect appropriate social values or shadow prices (for example, subsidized inputs such as fertilizers; subsidized seedlings provided in afforestation programmes; and labour inputs that would otherwise be unemployed).

3   Benefits and costs for which no market prices exist but for which appropriate social values can be approximated in money terms by inferring what consumers would be willing to pay for the product or service (for example, forest products, such as fuelwood or wild fruits, where market transactions are prevented by rules stating these resources are only for personal use; see Chapter 4).

4 Benefits and costs for which it would be difficult to imagine any kind of market-like process capable of registering a meaningful monetary valuation (for example, a sacred woodland).

The ease with which benefits and costs are estimated clearly diminishes as one moves down the preceding list. Benefits and costs in the last category are not commensurable and must be included as qualitative factors in the overall analysis. Of particular interest in the context of environmental decision-making is the third category – the case where non-market benefits or damages can be simulated or imputed. Economic theory has progressed furthest with the estimation of amenity services such as outdoor recreation or hunting. Some progress has been made in the estimation of the benefits of clean air and clean water. Much less progress has been made in estimating the functional benefits of a watershed or an ecosystem.

Problems in estimating benefits

For many projects, such as an afforestation programme with fast-growing trees, project outputs can be readily evaluated in terms of observed market prices. Where these prices are distorted away from social values derived from supply and demand, due to market failures or government intervention, they must be corrected to reflect true societal scarcity values or opportunity costs (benefits foregone in the best alternative use). Economists call these adjusted prices 'shadow prices' or 'efficiency prices'. For example, if a price-supported agricultural commodity had an observed price of $1 per unit but $0.25 was a direct or implicit government subsidy, the proper shadow price for that commodity would be $0.75.[6] Generally, in estimating total benefits or total WTP for

project output, one is estimating an appropriate area under a demand curve (see Box 4.1). If the output of the project increases supply and depresses prices, the analyst may evaluate increased output in terms of the average of the pre-project and post-project price.

The most serious benefit estimation problems, however, are those associated with non-market benefits (such as the value of outdoor recreation or of clear air or water) and with secondary benefits (secondary benefits could include, for example, the greatly expanded sawmilling operations that result from an afforestation programme). In all too many cost-benefit studies, the non-market benefits are not considered when they should be, whereas secondary benefits are included when they often should not be. Indeed, the problem of secondary benefits has been troublesome and controversial in developed countries (Veeman, 1978, 1985). Economists argue that secondary benefits can only be counted as benefits if resources are unemployed within the accounting jurisdiction and if the spin-off activity truly represents a net gain in economic activity within that area, rather than a shift in activities from one area to another. In developing countries, where labour resources are seldom fully employed but capital resources may be, the question of whether to include secondary benefits as legitimate benefits is more open. For their inclusion, the analyst must be convinced that the unemployment of labour and capital resources would be chronic over the life of the project. In general, economists remain reluctant to justify projects in terms of their secondary benefits.

The estimation of non-market benefits associated with environmental amenity and other services are of particular interest to ecologists and economists (see Chapter 4). Some project outputs or services – the provision of siltation reduction or environmental quality improvement are notable examples – have the technical characteristics of public goods, not private goods, such as fruit from exotic trees, or ploughs.[7] The estimation of non-market benefits involves the simulation of demand curves for recreation or amenity services through direct-survey techniques (where users are asked to reveal their maximum WTP for the services) or through indirect techniques such as the travel-cost technique (where WTP is imputed from travel cost, population and visitation information). Although non-market benefit estimation remains an imprecise art, it is imperative that such non-market values are considered in cost-benefit studies.

Problems in estimating costs

Most costs will be estimated in terms of actual or adjusted market prices for inputs. If part of the labour force involved in the project would be otherwise unemployed, the money costs of using these workers should be adjusted downward (in the extreme, to zero) to reflect the fact that their actual wage rate is greater than their true scarcity value to society. The other major problem that is apt to arise on the cost side (beyond the difficult problem of allocating joint costs to different uses when the project provides several uses or outputs) is the question of external costs. It is imperative that all uncompensated costs associated with (technological) external diseconomies are included in the social cost-benefit analysis (classic examples are pollution, downstream siltation and other forms of environmental degradation not incident on the economic agent generating the damage). Again, this is easier said than done because one is typically attempting to measure non-market values.

A special case where potentially large external costs may be imposed on society is associated with the loss and transformation of unique natural or scenic environments (Krutilla and Fisher, 1975). Sometimes society must make a mutually exclusive choice between preserving a unique wild area for its present and future amenity services or developing that area (eg through flooding it with a dam for irrigation or through mining development). The demand for using a unique natural area with very limited or no substitutes will be characterized by significant consumer surplus and the existence of considerable existence values. Great caution is dictated in irreversibly developing such unique phenomena of nature, whose amenity services are anticipated to increase in relative value over time.

Choosing a discount rate

Values associated with natural resources generally occur at different points in time. However, individuals are not usually indifferent to when a resource value occurs. If given the choice between receiving $100 now or $100 in ten years, individuals will generally choose to receive $100 now. Accordingly, we may assume that the $100 in ten years is worth less than the $100 now. Therefore, the $100 in ten years must be discounted to take into account the fact that it will not be available for some time. The problem of discounting, and the choice of an appropriate rate of discount, arise from the need to put all benefits and costs, regardless of their occurrence in time, on the same temporal footing.

Discount rates may be derived from capital markets where money is borrowed and loaned. Rates reflected in private markets reflect three components:

1 the real rate;
2 inflation; and
3 a risk premium.

The real rate represents the cost of borrowing money in a risk-free and zero-inflation environment. Real rates of interest tend to lie within the range of 1 per cent to 6 per cent. However, many market interest rates, especially in developing countries, are much higher than this. Assume, for example, that the real rate of interest is 4 per cent while the observed market rate is 35 per cent. If inflation were estimated at 26 per cent, then the interest rate would be 30 per cent in a risk-free environment. However, in our example, there is an extra 5 per cent added to the market rate because of risks associated with making the loan.

Typically, benefits and costs are projected ahead in real or constant dollar terms and discounted using a real interest rate. Real values are used in analyses because the effects of inflation affect interest rates and value estimates, thereby generally cancelling each other.[8] Discount rates used may or may not include risk factors. In one approach, a risk-free rate is used, so that risk can be factored in independently of the discount rate. Instead of increasing the discount rate to account for potential risk, the approach is frequently to concentrate on specific values that are uncertain, and then to conduct sensitivity analysis with respect to these specific values (see Box 5.2). An alternative approach is to increase the discount rate for more risky situations. Box 5.3 provides an example of this practice, combined with sensitivity analysis.

The above discussion so far has considered the different components of market, or private discount, rates. These are the discount rates that private firms face as they make their economic decisions in the face of market forces. These rates may be thought of as behavioural rates and are appropriate to use in assessing whether it is in a company's interest to participate in resource-development assessments, and for investigating household adoption of practices in household assessments. For example, Box 5.3 uses a 17 per cent rate for investigating private adoption by poor and risk-averse smallholders, while a 5 per cent rate is used in the social analysis. Reflecting the private and social accounting stances described above is the fact that there may be private and social discount rates. Private rates may be distorted from a social accounting stance, in that they do not represent the correct weighting of social interests. In other words, there may be problems with market-specified rates in that they do not correctly reflect how society values resources over time.[9]

In the face of market failures, it can be difficult to identify a social rate of discount.[10] Two main schools of thought predominate in the literature on choosing the social discount rate. Firstly, there is the opportunity-cost school. The logic here is that the social discount rate should reflect the rate of return that might be earned in alternative private or public projects. The second school of thought argues that the social discount rate should reflect social time preferences; this reflects how society, collectively, feels about present as opposed to future benefits.[11] In an imperfect world, it is difficult to reconcile these two approaches, and the eventual choice of a social discount rate is somewhat arbitrary. At the moment, in Zimbabwe, the appropriate real risk-free rate could range between 2 per cent and 10 per cent, with lower values reflecting the time-preference orientation and higher values the opportunity-cost orientation. Because of its importance and the unresolved issues relating to the proper method of its determination, cost-benefit studies should generally include sensitivity analysis with respect to the discount rate.

Selecting among decision criteria

It is important to recall that each of the three types of assessments discussed above addresses different, but sometimes similar, types of questions. For example, resource development and household assessments may attempt to determine whether an activity or project is feasible for, respectively, a government (on behalf of society) or a household. Answers to these types of questions may vary according to the criteria that are used to summarize findings of assessments. Criteria used in practice generally concentrate on utilizing the benefit and cost information that has been collected, discounted over time. Accordingly, three criteria are generally considered (Dasgupta and Pearce, 1972).[12]

Firstly, there is the measure of net present value (NPV). Net present value is defined as the difference between discounted benefits and discounted costs. Accordingly, if NPV > 0, then time-adjusted benefits exceed time-adjusted costs. In the context of development-project assessments, a NPV > 0 implies that a project is feasible. Furthermore, among those set of feasible projects, alternatives with higher NPVs are ranked higher than competing projects with lower NPVs. Similarly, if a number of variants of a project are possible, the variant with the highest NPV is considered the best variant. In the context of green-accounting assessments, NPVs of resource endowments that are stable or that increase over time are preferred to NPVs that decrease over time. Finally, in the context of household assessments, activities with higher NPVs are more likely to be undertaken than projects with lower NPVs, because higher NPV projects are assumed to contribute more to livelihoods.[13] Box 5.2 illustrates the use of NPV.

A second measure, closely related to the first, is the benefit-cost (BC) ratio. Benefit-cost ratios are calculated by dividing total benefits by total costs, where benefits and costs have been adjusted for time (ie discounted). Accordingly, project feasibility is determined by a ratio exceeding the value of one. Similarly, projects with higher BC ratios are preferred to those with lower BC ratios. The key difference between NPVs and BC ratios lies in the way that differences between benefits and costs are expressed. Whereas NPV compares benefits and costs in absolute terms, BC ratios compare benefits and costs in relative terms. Box 5.3 uses a BC ratio to evaluate the viability of planting a medicinal plant.

The third measure is termed an internal rate of return (IRR). The IRR of a project describes that interest rate which makes a NPV = 0 or a BC ratio = 1. Rather than expressing the benefits above costs in absolute or relative terms as is done, respectively, for NPVs and BC ratios, the IRR expresses the potential for having benefits above costs as a percentage. Whether the IRR percentage represents benefits greater than costs depends upon the specific benefit and cost estimates, when they exist in time, and the discount rate to which the IRR is compared. For example, assume that we are assessing a development project from a social perspective and the project yields an IRR of 10 per cent. The analyst then compares the IRR to what she or he perceives is the appropriate social discount rate. If the social discount rate is perceived as being > 10 per cent, then the project has costs greater than benefits, and the project fails. If the social discount rate is perceived as being < 10 per cent, then the project has benefits greater than costs, and the project is considered feasible. In terms of ranking, projects with higher IRRs are considered superior to projects with lower IRRs. Box 5.5 illustrates the use of IRR in the financial analysis of livestock and wildlife enterprises.

These three criteria, and elements within these criteria, are often used in various ways. For example, NPVs and BC ratios may be expressed on a project or scenario basis (eg Boxes 5.2 and 5.3, respectively), or on a per hectare or holding basis (eg Boxes 1.4, 2.2, 5.1 and 7.8). In other cases, returns may be reported relative to some physical measure of input, rather than relative to costs of all used inputs. For example, in Box 5.1 benefits are expressed per unit of labour and capital. Such measures may be useful in situations where the benefits are known but the values for the inputs are not. Furthermore, such measures may be particularly insightful in situations where there is one input that is particularly scarce locally, and therefore represents an important constraint on local income generation. In cases where values of benefits and costs are known, measures such as BC ratios may be preferable because they provide an overall assessment of dollars received per dollar invested. Finally, in some cases the above criteria may be used for estimates representing a single year (eg Box 5.1). In other cases, criteria may be presented for a single year and then extrapolated over time (eg Box 5.6). In cases of single years, it is not necessary to use discounted values, since values are all measured at one point in time. However, where values are taken over multiple years, discounted values should be used.

Although it is beyond the scope of this chapter to describe, in detail, the use of these three criteria, a few problems should be noted. Firstly, although all of these criteria will yield the same results for questions regarding feasibility, differential rankings can be obtained depending upon what criterion is used. In general, the NPV criterion is preferred over the BC ratio or IRR for ranking purposes.

Furthermore, with the IRR, it is possible to have multiple solutions when income streams with alternating benefits and costs occur. Accordingly, it is advisable to become more familiar with these criteria before they are used (see, for example, Gunter and Haney, 1984).

The three criteria, discussed above, concentrate on using benefit and cost information, adjusted for their places in time, in order to make decisions. These types of measures are frequently referred to as measures of economic efficiency. Historically, cost-benefit analysis has focused on the criterion of efficiency – that is, the relationship of benefits and costs exclusive of their distribution. Ideally, economists would prefer to evaluate projects or policy in terms of both efficiency and equity. Any project with net benefits greater than zero (or, alternatively, benefit-cost ratio exceeding unity) passes the limited test of economic feasibility. A more stringent efficiency test requires that the project be compared with alternative projects or uses of scarce investment funds. Any project will also have impacts on the personal, regional and functional distributions of income. Economists are increasingly attempting to bring these distributional aspects into the traditional cost-benefit format by either attaching distributional weights to various classes of benefit receivers or, alternatively, by adding a qualitative statement to the end of the standard-efficiency analysis, which examines who reaps the benefits and who bears the costs. Similarly, environmental objectives and values are being incorporated into cost-benefit studies. If environmental benefits and costs can be measured or simulated, they are included directly in the study; otherwise, they are considered qualitatively.

BOX 5.6 IS THERE AN ADVANTAGE TO CULTIVATING PALMS FOR BASKETRY?

This box describes a study that compared basket-making based on wild populations of palm with that based on cultivated palm populations (Bishop and Scoones, 1994; see Figure 5.3). The data come from two villages in Botswana (see also Box 5.4). Potential time savings in collection are assumed to be the main advantage of cultivating the *Mokola* palm as a substitute for wild material. The case study demonstrates the numerous pieces of information that are needed in order to perform the relatively simple economic comparison.

Labour for basket-making

Current labour inputs to collect raw materials are highly variable, depending upon the proximity and accessibility of desired plant species. Considerable time is used getting to and from the main collection sites. Moreover, the time required to gather raw materials varies depending upon their relative abundance at a given location and the methods used to collect them (eg whether a knife or axe is used to cut palm leaf). Data on travel and collection times for different sites were obtained through direct observation by accompanying women on collection outings, as well as recall estimates (see Table 5.7). An estimate of the number of N5 baskets (typically, open baskets of 37cm in diameter by 8cm in height) that may be produced from one head load was obtained by key informant interviews, using samples of raw and processed palm leaf and a N5 basket as an aid to memory.

Table 5.7 *Time required to collect bundles and output of baskets per year*

Interview	Travel time (hours)	Bundles per trip*	Bundles required to make one N5 basket	Total trips per year per person	N5 baskets that can be made from the material collected on one trip	Baskets produced per year per person
Average	10.9	10.1	2.6	14.1	4	55.3
Minimum	8	6	2	10	2	30
Maximum	13	15	3	20	5	85

* The number of processed bundles of fibres that results after processing one head load of material (women collect a single head load of leaves during each trip)

Labour inputs for processing vary depending upon whether the palm leaf is dyed or left white. In the former case, and assuming that the raw dye material has already been obtained, a total of 3.5 hours is required to pound the dye into powder, remove the thick edge of each palm leaf segment and then boil the powder and leaf together. When the palm leaf is left white, roughly one hour is required to split and boil one head load. Total labour inputs for processing the raw materials that go into one N5 basket thus depend upon the relative proportion of dyed leaf used in an average basket. Interviews with weavers suggest that the proportion of dyed to white leaf used is about 1:1 in Etsha and approximately 2:7 in Danenga. Assuming an average of four N5 baskets may be produced from a single head load (see Table 5.7), a range of estimates was obtained of the total time spent processing raw materials per N5 basket (see Table 5.8).

Weaving was estimated to take approximately 20 hours per basket. Collection times are, therefore, a rather small part of the total labour inputs to basket-making (see Table 5.4).

Table 5.8 *Labour inputs for processing raw materials*

Location	Ratio of white to dyed leaf	Total labour to process one head load (hours)	Total labour to process the raw materials for one N5 basket
Etsha	1:1	2.50	0.63
Danenga	2:7	3.14	0.79

Tools

Different tools are required for each step of production (see Table 5.9). The cost of the tools is calculated as a function of the purchase price, useful lifetime and share of use in basket-making (versus other domestic uses). For tools used in collection, the cost is calculated as an average cost per head load (trip), which is then converted into a cost per basket, knowing how many baskets are produced from each head load. For processing tools (excluding a knife that is already valued), the average cost is calculated per N5 basket. The average cost of tools for all production stages is insignificant, at only 0.18 pula per basket (less than 3 per cent of the producer price) – a feature of many NTFP enterprises.

Table 5.9 *Tools for collecting (c) and processing (p) raw materials, and for weaving (w)*

Tools	Cost in 1993 (pula)	Useful life (years)	Share of use for basket-making (%)	Charge per head load or trip (pula)	Charge per N5 basket (pula)
Axe (c)	20	2	25	0.18	0.05
Hoe (c)	20	1	5	0.07	0.01
Knife (c, p, w)	5	1	50	0.18	0.05
Sickle (c)	4	5	50	0.03	0.01
Awl (p)	0.15	7	100		0.00
Mortar and pestle (p)	60	3	5		0.03
Pot (p)	30	10	5		0.01
Blades* (w)					0.03
Total					0.18

* One blade is used per basket.

Selling prices

Prices paid to weavers vary with the size and shape of the basket. Using data on the proportion of baskets in different grades, and the prices for those grades, gives a weighted average price of 7.85 pula for one of the purchasing cooperatives and 9.52 pula for another.

Returns to labour

If one subtracts tool costs from the weighted-average selling prices paid, one obtains the gross margin per N5 basket (net of tool costs): 7.66 pula in Etsha and 9.54 pula in Danenga. Gross margins are divided by total labour inputs per basket in order to obtain an estimate of returns to labour on an hourly basis (see Table 5.5). These returns thus incorporate resource rents.

Cost-benefit analysis of palm cultivation

With the increasing scarcity of suitable wild palm near settlements and the greater distance that weavers are forced to go to collect it, there is growing interest among the main buying agencies in promoting the cultivation of *Mokola* palm. The potential benefits of cultivation can be expressed in terms of reduced travel times. Table 5.10 indicates the minimum and maximum returns to labour (pula per hour), using cultivated and wild palm, in Etsha and Danenga. A range of values is given to account for different assumptions about the time required to collect palm leaf and the amount of material obtained during one trip. The difference in returns to labour from wild and cultivated palms is a measure of the benefit of reducing travel time to zero, assuming that cultivated palm is immediately accessible in gardens located near the home. As shown in Table 5.10, the returns to labour through the use of cultivated palm are only a maximum of 24 per cent higher than the returns to labour through the use of wild resources. Under less optimistic scenarios, the percentage drops to 4 per cent. This reflects the relatively modest share of travel time in total labour inputs for basket-making.

Table 5.10 *Returns to labour from wild and cultivated palms*

Source of palm	Etsha		Danenga	
	Minimum pula/hour	Maximum pula/hour	Minimum pula/hour	Maximum pula/hour
Cultivated	0.37	0.35	0.46	0.42
Wild	0.34	0.28	0.44	0.35
Difference (equals benefit of cultivation)	0.03	0.07	0.02	0.06
% difference	9%	24%	4%	17%

On the cost side, palm cultivation necessitates fences in order to protect the young plants from livestock. *Mokola* palm matures quickly; in plantation it is assumed to produce leaves suitable for making baskets from the third year. Estimates were available for supplying and installing fencing.

Table 5.11 shows the total labour savings obtained through cultivation over a ten-year period. Labour savings represent travel hours saved from not having to collect wild palm (actual harvest time remains the same). There are no savings in the first two years, while the palm matures; but from the third year it is assumed that all leaf is obtained from cultivated palm. Note that labour savings are discounted (at 5 per cent) to reflect the fact that benefits received in later years are 'worth' less, from today's perspective.

Table 5.11 *Cumulative labour savings from cultivating palm (assuming minimum and maximum values*) and cost per hour saved*

Year	Discount factor (r = 5%)**	Etsha Minimum discounted hours/year	Etsha Maximum discounted hours/year	Danenga Minimum discounted hours/year	Danenga Maximum discounted hours/year
1	1.00	0	0	0	0
2	0.95	0	0	0	0
3	0.91	17	95	7	73
4	0.86	16	91	7	69
5	0.82	15	86	7	66
6	0.78	15	82	6	63
7	0.75	14	18	6	60
8	0.71	13	15	6	57
9	0.68	13	71	5	54
10	0.64	12	68	5	52
Total discounted hours saved		115	646	49	492
Cost (pula per hour saved)***		0.48	0.09	1.12	0.11

* Minimum hours saved assumes minimum travel time, maximum baskets per head load and minimum trips (baskets) per year. Maximum hours saved assumes the opposite.

** The discount rate appears to be the wrong rate to use. The rate should be a private rate, which presumably would be much higher.

*** Cost per (discounted) hour saved assumes a capital cost of 55 pula per household for fencing.

Finally, one can compare the potential benefits of cultivation – expressed in terms of increased returns to labour from reducing travel time to zero – to the costs of cultivation – expressed in terms of the cost of fencing per discounted hour saved. This comparison is shown in Table 5.12 and indicates that the costs of cultivation exceed the potential benefits in every case. Note that the additional labour costs of cultivation have been deliberately excluded from the analysis; adding these costs would further undermine the case for cultivation.

Table 5.12 *Cost-benefit analysis of palm cultivation*

	Etsha Minimum pula/hour	Etsha Maximum pula/hour	Danenga Minimum pula/hour	Danenga Maximum pula/hour
Benefit (Table 5.10)	0.03	0.07	0.02	0.06
Cost (Table 5.11)	0.48	0.09	1.12	0.11
Net benefit (B – C)	(0.45)	(0.02)	(1.10)	(0.05)

Note, however, that Table 5.12 also indicates that the costs and benefits of cultivation begin to converge under the 'maximum hours' scenario – in other words, when maximum travel times and minimum numbers of baskets per trip are assumed (ie per head load of raw material). Only with much greater resource depletion and much higher travel times will cultivation become an option.

Note: See 'Surveys of labour' in Chapter 2 for possible methods, and Box 5.6 for an illustration of incorporating labour costs into a NPV analysis.
Photograph by Manuel Ruiz Pérez

Figure 5.2 *Preparation of condiment from the bark of* Scorodophoeus zenkeri *in Cameroon. Many of the techniques in this chapter require that the labour costs are taken into consideration*

Photograph by Tony Cunningham

Figure 5.3 *A* Hyphaene petersiana *palm planted by a basket-maker in north-western Botswana*

Cautions in using benefit and cost information

To what extent can the analysis of benefits and costs play a useful role in making wiser decisions about resource and environmental use? There can be no doubt that the general cost-benefit approach of assessing the advantages and disadvantages of a particular course of action, policy or project must lie at the heart of any sensible approach to public decision-making, including that related to natural resources and rural livelihoods.

There is a danger, however, in becoming too blinkered by rose-colored glasses as one extols the virtues of frameworks using cost-benefit information. Unfortunately, cost-benefit analysis can be abused, misused or be found grossly deficient in the assess-

ment of public policy or projects. To some, cost-benefit analysis is a precise 'science' that is capable of rendering a specific benefit-cost ratio and 'clear-cut' guide to public policy. In fact, cost-benefit analysis is as much an 'art' that typically rests on considerable judgement and seasoned intuition on the part of the analyst. All too often, cost-benefit analyses have involved incomplete or improper specification of the problem at hand, the overstatement of project benefits (eg the padding of the benefit ledger with secondary benefits) or the understatement of project costs (due to cost overruns or the omission of environmental damage). Indeed, it is possible to become quite cynical about cost-benefit analysis whenever it is converted to play the role of legitimating political decision or, more harshly, of being the 'lubricant of politically sanctioned greed' (Bromley, 1980). Clearly, an analytical tool may be used or abused by political interests.

Last but not least, economists recognize several theoretical and practical problems with cost-benefit analysis. Firstly there are conceptual difficulties with a policy approach that rests on debatable foundations of applied welfare economics. In the absence of more developed measures of welfare, the typical focus of cost-benefit analysis is on the dollar values of benefits and costs, with little or no attention to questions of equity or income distribution. There is also an emphasis on WTP as the basis for benefit estimation when WTP, like other measures of welfare, is conditioned on the current distribution of income and ownership of resources in society.[14] It is difficult to incorporate the preferences of future generations and, more generally, to deal with policy problems that have planning horizons exceeding 30 or 40 years. Discounting means that costs and benefits from the distant future have minimal value. Cost-benefit studies involv-

ing rare and endangered species will need to be supplemented with policy recommendations based on the logic of safe minimum standards or the precautionary principle. In addition, the data requirements of cost-benefit work are often very high. This is illustrated in Box 5.6, which represents a rather simple comparison of whether it would be viable for households to plant palms rather than collect leaves from the wild. In spite of the simplicity of the comparison, a wide range of variables needs assessment and estimation.

Cost-benefit analysis may not be perfect or even fully perfectible. But cost-benefit logic in assessing the pros and cons of alternative courses of action remains unassailable as a general rational approach to public policy. Gradually, too, some (though not all) of the difficulties associated with cost-benefit analysis are being overcome – for example, the greater focus on, and greater sophistication in, evaluating environmental and distributional considerations. Finally, one cannot give up on a potentially useful policy tool merely because it has been, or is being, subverted or ignored by planners or politicians.

Notes

1   Although there are increasing efforts in 'green' accounting, such methods are not yet prevalent as a part of national accounts, largely because of a lack of information regarding natural resource stocks, flows and associated values.

2   The unit of analysis may be individuals, households or larger groupings of individuals. For a discussion of these issues, see the section on 'The household as the unit of analysis and handling absentee members' in Chapter 2 and the section on 'Influences of social differentiation within households on values and behaviour' in Chapter 8.

3   The World Bank, a leading institution that undertakes project analysis in the world, differentiates between commercial project appraisal, economic project appraisal and social appraisal. *Commercial project appraisals* include private benefits and costs, received, and borne, by firms. This term is used interchangeably with *private appraisals* (in Box 5.3) and *financial appraisals* (in Box 5.5). *Economic project appraisal* (traditionally called social project appraisal) refers to appraisal done in terms of society's benefits, costs and discount rate (including external benefits and costs and correction of any distortions in the market

prices of outputs and inputs). The term *social appraisal* is now reserved by the World Bank to refer to appraisal that reflects consideration of income distribution as well as efficiency objectives. However, in Box 5.3, the term social appraisal refers to an economic project appraisal.

4   Market prices may be distorted due to market failures, such as external costs and benefits mentioned above, or due to government policies that may subsidize or tax goods and services. In these cases, it may be necessary to estimate shadow prices that represent the true resource values (see Box 5.5 for example). A considerable literature has arisen concerning appropriate shadow prices for labour and for foreign exchange in developing countries (see, for example, UNIDO, 1972).

5   There are a number of regulatory possibilities that may be used to influence household incentives and behaviour. Some of these are discussed further in Chapter 8.

6   See Note 4.

7   A public good has the technical characteristic that its benefits are jointly shared by all users (up to the point of congestion or capacity limitation); a related, and likely derivative, characteristic is that it is impossible (for reasons of technology, cost or economic efficiency) to exclude a second user once the service has been provided to the first user.

8   In cases of hyperinflation, the effects of inflation do not cancel out exactly, creating the potential need to use nominal interest rates and nominal benefits and costs.

9   A further problem with market-determined interest rates is that they are subject to market fluctuations that may vary over the life of a project.

10 For a review of issues regarding the social rate of discount, see Luckert and Adamowicz (1993), Hanley (1992), or Dasgupta and Pearce (1972).

11 One issue associated with this approach is how to estimate what this rate might be.

12 Other criteria may also be used. For example, Box 5.5 makes use of the domestic resource cost (DRC) ratio. Variants on the three criteria discussed below are presented in the latter part of this section.

13 Net present values are only one consideration in adoption behaviour. A number of other factors, including risk factors, are discussed in Chapter 8.

14 For issues associated with WTP as a representation of social values, see Chapter 4.

Chapter 6

Participatory methods for exploring livelihood values derived from forests: potential and limitations

Nontokozo Nemarundwe and Michael Richards

Introduction

Since the 1980s, rural development research has gradually shifted from the use of conventional extractive approaches towards participatory investigation and analysis (IIED, 1997). Participatory approaches, such as participatory rural appraisal (PRA) (see Figure 6.1), have been used both as development and research tools. The emphasis here is on the use of PRA as a research tool. In previous volumes in this *People and Plants* series, PRA approaches have been introduced (Martin, 1995; Tuxill and Nabhan, 2001). This chapter describes the procedures for planning and conducting PRA to enhance interactive participation by local communities in the process of learning about rural people's values, with regard to trees and forests. Field experiences show that there are various potential benefits for the people who adopt the PRA approach in the research process. PRA gives researchers the opportunity and skills to facilitate local people to articulate their opinions, identify and prioritize their problems and needs and, most importantly, to seek ways and means of solving their problems.

Over the years, PRA has confirmed that local community members are more knowledgeable than they are often portrayed to be. Given appropriate attitudes and behaviour, patience and skills, development practitioners and researchers quickly learn that there is knowledge at the grassroots level, although it is of a different form and nature from that to which they are accustomed. With PRA, community knowledge and values can be identified and integrated with data derived from other research methods.

PRA tools also have a number of serious drawbacks, particularly if PRA scoring techniques are used to quantify forest values. This chapter highlights and discusses these drawbacks, concluding that PRA and more conventional tools can play complementary roles, and that for rigorous quantification, techniques such as those outlined in Chapter 2 and 4 are necessary.

The first part of this chapter looks at the concept of 'participation'. The section on 'The background, history and principles of PRA' describes PRA, giving a brief

Photograph by Bruce Campbell

Figure 6.1 *A role-play conducted by three women amidst a PRA group in Hot Springs, Zimbabwe. Role-plays were used to identify the range of woodland benefits*

background and identifying the main principles. This is followed by the section on 'PRA tools and techniques for quantifying and valuing forest benefits' that describes some of the PRA tools that can be used for resource valuation and for assessing the economic contribution of

trees and forests to livelihoods. Finally, this chapter concludes with a discussion of the strengths and weaknesses of PRA ('Strengths and weaknesses of PRA for understanding forest values') and some proposals for the way forward.

What is participation?

Because the term 'participation' is so widely used in research and development literature and so variously interpreted, there is a need for an operational definition. The meaning of participation can range from almost complete outside control, with token involvement of local people, to a form of collective action in

which local people set and implement their own agenda in the absence of outside initiators and facilitators (Carter, 1996). Between these two extremes are various intermediate forms of participation that are illustrated in Figure 6.2.

The bottom step of the participation ladder is referred to as passive participa-

169

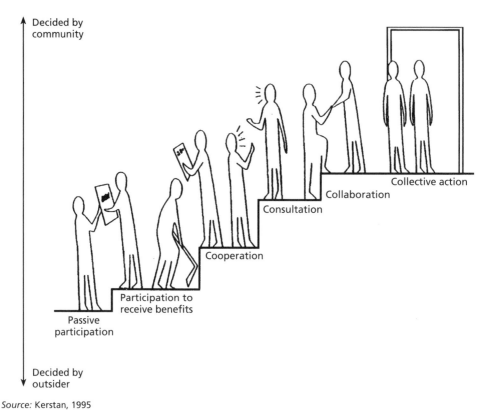

Decided by
community

Passive
participation

Participation to
receive benefits

Cooperation

Consultation

Collaboration

Collective action

Decided by
outsider

Source: Kerstan, 1995

Figure 6.2 *The ladder of participation*

tion. Pimbert and Pretty (1995) describe passive participation as the situation where people 'participate' by being told what is going to happen or what has already happened. People's responses to what they have been told are not taken into account. There is no information sharing between people and the external facilitators. The next step is participation to receive benefits (often, material benefits). In this case, people participate as long as there are benefits to be realized. Once the benefits run out, then they have no stake in prolonging the activities. An example from Zimbabwe of this type of participation is the programme commonly known as 'food for work'. Under this programme, villagers were asked to implement development activities in their villages, such as road

rehabilitation and gully reclamation, in exchange for food. Once food was no longer available, people stopped participating, even in cases where the projects were not yet complete. This form of participation often creates some dependency syndrome on the part of the villagers.

The third step in the ladder of participation is what Carter (1996, p4) has referred to as cooperation. In this case, people participate to take action prescribed by the external facilitators. The external professionals decide the agenda and direct the process. People tend to depend upon the external facilitators, but given a chance, they could be self-dependent. The fourth step up the ladder is consultation. In this case, the outsiders decide the course of action to take, but after soliciting people's opinions. The outsiders define the problem and are under no obligation to take into consideration the people's views. The fifth step is participation through collaboration. In this case, local people work together with external facilitators in order to determine priorities. The external facilitators have the responsibility of directing the process. The sixth step, one that development practitioners who aim to be 'genuinely' participatory should work towards, is referred to by Pimbert and Pretty (1995) as interactive participation or collective action. In this case, the local people participate in joint-problem analysis, which leads to action plans. This process tends to involve interdisciplinary methodologies that seek multiple perspectives and make use of systematic and structured learning

processes (Pimbert and Pretty, 1995). The people take control over local decisions so that they have a stake in maintaining structures or activities.

Many of these steps in the 'participation ladder' apply to development initiatives and are less easily applied to research projects. Using the definitions set out above, most 'participatory' forest-resource assessment falls in the categories of consultation and/or collaboration. It is worth noting that many participatory forest-management initiatives do not fall exclusively into the various steps described above. They may have characteristics of one or more of the various forms of participation. PRA aims at achieving interactive participation, although this is very difficult, especially for research projects that do not have a tangible development agenda.

The background, history and principles of PRA

The past decade has witnessed a number of shifts in the rhetoric of rural development research. These shifts include the now familiar changes from 'top down' to 'bottom up', from centralized standardization to local diversity, from conventional research to learning processes. Linked to these have been small changes in modes of learning, from extractive research modes to participatory appraisal and analysis. In these changes, a significant part has been played by PRA. Using PRA in the research process means increasing the space for participants to express themselves and control the knowledge that is being created (Goebel, 1998). The researcher limits the imposition of analytical categories in the data-collection process. While it has been argued that academic research will remain essentially extractive in nature, a researcher using PRA is normally trying to promote empowerment.

PRA has been defined as 'a family of approaches, methods and behaviours that enable people to express and analyse the realities of their lives and conditions, to plan themselves what action to take, and to monitor and evaluate the results' (Chambers, 1992). PRA is both a philosophy (outsiders need to learn about situations from insiders, and insiders can analyse their own problems) and a series of methods for carrying out participatory and qualitative research. PRA emphasizes processes that empower local people.

The principal aim of the PRA process is to give more power to the community and to reverse power relations and hierarchies between communities and those perceived as being development experts and planners from outside (Chambers, 1992). The approach argues that development programmes should be designed and controlled by the communities affected,

Table 6.1 *A comparison of some of the features of RRA and PRA*

Features	RRA	PRA
Period of major development	Late 1970s and 1980s	Late 1980s and 1990s
Organizations responsible for the major innovations in the methodology	Universities	Non-governmental organizations (NGOs)
Main users	Aid agencies; universities	NGOs, government departments and universities
Key resources that the approach recognizes and that were earlier overlooked	Local people's knowledge	Local people's capabilities and capacities
Main innovations	Methods	Behaviour
Predominant mode	Elective, extractive	Facilitating, participatory
Ideal objectives	Learning by outsiders	Empowerment of local people
Long-term outcomes	Plans, projects and publications	Sustainable local action and institutions

Source: Chambers, 1992

rather than by external agencies. Emphasis is on interactive processes between external agencies and local communities.

Background and history

PRA evolved partly from rapid rural appraisal (RRA). RRA had four main origins. The first came from the dissatisfaction with biases, especially the biases of rural development tourism – the phenomena of brief visits by urban-based professionals. The second origin of RRA came from the disillusionment with questionnaire surveys and their results (see Chapter 2 for details on questionnaire-based approaches). Experience with questionnaire surveys suggested that they were tedious for respondents and unlikely to aid insight into many important phenomena (eg the micro-political processes driving differentiation, the reasons for certain components of the landscape or activities that were highly valued). The third origin was the growing recognition by development practitioners of the painfully obvious fact that rural people were themselves knowledgeable on many subjects that touched their lives.

What has become known as indigenous technical knowledge (ITK) was increasingly seen to have richness and value for practical purposes (Chambers, 1992). The fourth origin was that RRA could be implemented cheaply and quickly, unlike the more costly and time-consuming questionnaire surveys, thus attracting the attention of development agencies. The section on 'Strengths and weaknesses of PRA for understanding forest values' returns to some of these issues when discussing the weaknesses of PRA techniques. PRA can, in fact, be costly to local people in terms of time, and PRA is often implemented rapidly to the point where it differs little from rural development tourism.

The most common mode of RRA entails outsiders obtaining information, taking it away and analysing it. This practice is extractive because the information is not shared and owned by local people. It is the recognition of this weakness within RRA, and the need to enable people to analyse and modify their own situation, that have led to the transformation of RRA into PRA. A summary comparison of what features are normally described as RRA and PRA is given in Table 6.1. In terms of research, much of what is described as PRA is more truthfully RRA. It remains largely extractive and rapid. Longer-term engagement with communities during research is found in some action-research projects (Long and Long, 1992).

Features and principles

Of the principles listed below many are shared with RRA (Chambers and Guijt, 1995). Essentially, the key features of PRA include sequencing, joint learning and an emphasis on facilitation skills.

Multidisciplinary teams

The principle behind the use of a multidisciplinary or interdisciplinary team is that people with different skills, experience, perspectives and points of view will look for and find different things, thus obtaining new and deeper insights (Martin, 1995). As discussed in Chapter 7, this is not always easy to accomplish. All members of the team should be involved in research design, data collection and analysis (see the section on 'Partnerships, interdisciplinary research workshops and teamwork' in Chapter 7). The interdisciplinary approach promotes the sharing of ideas, experiences and a better understanding of the diversified way of life in the given communities. PRA is a learning experience in which the team members learn from each other and from the community.

Selecting and sequencing the tools

Plans and research methods are semi-structured and are revised, adapted and modified as the PRA fieldwork proceeds. This is a process in which the study team constantly reviews and analyses its findings in the field in order to determine

in which direction to proceed. It builds up understanding and narrows the focus of the PRA as it accumulates knowledge.

The selected tools are taken from a wide range of possible tools that are tailored to the specific requirements of the study. Using different techniques gives greater depth to the information collected, improves the quality of data collection and promotes triangulation (see below).

As a research process, PRA moves through a logical sequence of tools. For instance, the facilitator can start with mapping. In some cases, a historical profile of the community can be the first activity. Care must be exercised in selecting the opening tool as it can determine the relative success of the PRA. For instance, in Matabeleland South in Zimbabwe, a research team investigating issues concerning malnutrition in children under the age of five found that starting with a historical profile was very sensitive, because of the past civil war in the area. Later in the PRA, villagers refused to draw a social map because they thought this might be used to locate their homes and get them killed. Similarly, in Hwedza District in Zimbabwe, a foreign researcher who began with a wealth-ranking exercise found that this could spoil the whole meeting. Some participants later refused to express their views, and suggested to the researcher that because they had been ranked as poor they knew nothing. The choice and sequence of tools and techniques should be adapted to fit each situation. A sequencing of tools used in one valuation exercise is shown in Figure 6.3.

Interacting with the community

PRA espouses the view that there has to be a culture of sharing information and ideas between insiders and outsiders, as well as sharing experiences between differ-

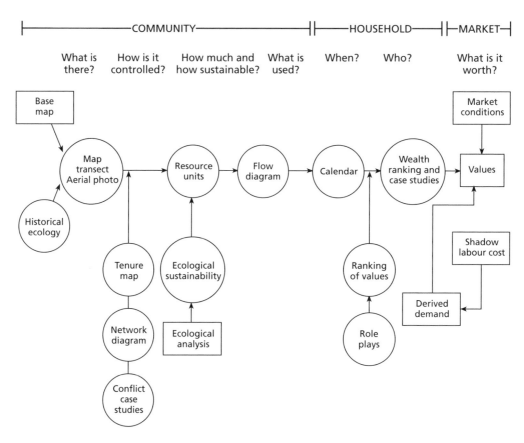

Figure 6.3 *The proposed sequence of methods used to value resources in the Hot Springs area. The PRA methods are shown in circles while other methods are shown in squares. The questions asked in the research process are also shown*

ent organizations. Of course, for inexperienced researchers who are initiating work in communities, the level of information that they can give to local people is not very high. Similarly, in short-term research projects joint learning may be difficult to achieve. The intention in PRA is that outsiders should interact closely with the community members and be able to see the lives and problems of the community through their eyes. Most activities are performed jointly with the community members, or by them on their own. Involving community members facilitates interpretation and understanding of the data collected.

Facilitating investigation, analysis, presentation and learning by the people themselves, so that they present and own the outcomes, and also learn, has been termed 'handing over the stick'. This entails an outsider starting a process and then sitting back and not interrupting. While this may be feasible when using PRA as a development tool, it is seldom used to any great extent in the research process.

The emphasis of PRA on group-based discussions demands good facilitation skills. In the absence of this, the exercises are unlikely to be fruitful. As described by IIED (1997), a good facilitator is one who will ensure that as many voices as possible are heard in a group setting and that the findings represent the majority of opinions. It is the behaviour and attitude of the external facilitator, more so than the tools used, which contribute to the final success of the PRA process. The important attitudes include: critical self-awareness; embracing error; sitting down; listening and learning; not lecturing but handing over responsibility to the villagers; having confidence that the villagers can conduct the exercise; and being relaxed and open. There is a need to be culturally sensitive. PRA is a process; thus, facilitators must be patient. During the PRA research process, emphasis should be placed on establishing a more equitable relationship between the 'outside' researchers and the local people.

Attitudes to data

During data collection, trade-offs need to be made between the quantity, relevance, accuracy and timeliness of the data. The principle of optimal ignorance needs to be considered – knowing what is worth knowing. Another relevant principle is that of appropriate imprecision – not measuring more than is needed. In short, it is better to be approximately right than precisely wrong. Biases must be offset, requiring the researcher to constantly examine the data collected. Thus, possible sources of error must be identified and considered in the light of how they influence the way information is interpreted. This requires obtaining views from a cross-section of society, including the least assertive members of the community (the poorest, people with disabilities, women, children, ethnic minorities, the uneducated and those living in the most remote areas).

Triangulation of the data is necessary. This involves cross-checking qualitative information using a range of methods, varied information, different investigators, and/or a multitude of disciplines (Chambers, 1992). In order to be reliable, information about the same thing should be collected in different ways from at least three sources.

PRA tools and techniques for quantifying and valuing forest benefits

The choice of tools varies according to the particular issues being investigated and the local context. New tools are continuously being created and added. PRA practitioners are encouraged to experiment; there is not just one way of doing it 'right'. PRA does not offer a fixed methodology but offers a basket of tools for use within the given context.

When planning a PRA activity, it is advisable to first list the issues that need to be investigated and then describe the techniques that can be used to tackle each issue. There are many PRA manuals available that show the full spectrum of tools (eg Pretty et al, 1995; Messerschmidt, 1995). This chapter focuses on the PRA tools that can be used in order to learn about how people value forests, including:

- spatial tools (resource mapping, social mapping);
- role-playing;
- ranking and scoring, such as wealth-ranking (eg Martin, 1995, section 6.4.1), pairwise-ranking, simple ranking/scoring, matrix-ranking/scoring, contingent-ranking/scoring;
- the barter-exchange approach.

For a longer list of possible techniques that were used in one valuation exercise, see Figure 6.3.

Mapping forest resources

Resource mapping (see Box 6.1 and Figure 6.4) may be used as a starting point for village group discussions and can help to identify the range of tree and forest products. This information is useful for later ranking and scoring exercises. In the Zimbabwean context, mapping exercises are good as starting points at PRA workshops. Resource maps give a visual impression of the spatial situation that relates to the village or individual household. The maps can show where resources, activities problems or opportunities are located. They also help in assessing the dimension and scope of the issues to be investigated. Resource maps that are drawn by community members can be used with other more 'formal' tools, such as aerial photographs, survey maps and geographic information system (GIS) maps (Powell, 1998). Analysis of these 'formal' tools can also be done in a participatory manner. Information generated may reflect relationships such as the allocation of land to various agricultural activities, where people obtain forest products such as firewood and fruits, or where forests are used for traditional ceremonies. Discussions centred on the maps can also be used to explore the context of the resources to be valued, the rules and regulations pertaining to the resource units (see Figure 7.5) and conflicts over resources.

Role-playing

Role-playing can be used to identify resources that are valued, and to explore conflicts over resources. This involves

---

Box 6.1 Undertaking a resource-mapping exercise

A team comprised of researchers, extension field staff and community representatives should ideally undertake such an exercise. The various parties bring different but complementary ideas to the process. A first step is to ask community members to draw a map of their village and superimpose the features of main emphasis. Participants should be supplied with coloured pens and large sheets of paper, or asked to draw on the ground with sticks using leaves, stones and sand to show the various features. Some symbols (eg berries) can also be used to show the distribution of particular forest resources (eg fruit-collecting areas). It is important to 'interview' the map so that participants are encouraged to show more features and to openly discuss the mapped elements. It is important not to intimidate the community by using sophisticated aids.

The activity can be an exciting introduction to the issues, although it is time consuming, often taking up to two or three hours. It is important that it is not rushed, as the information generated can be very valuable – in particular, the open-ended discussions that occur.

Figure 6.4 illustrates one map that was produced in Jinga village, Zimbabwe. Apart from setting the context for the valuation study that followed, the map was used to identify the resource areas from which products were obtained. These resource units were subsequently evaluated for their institutional arrangements (see Figure 7.5) and for the benefits that each unit provided (see Figure 7.6). The map was also used for the random selection of households for the questionnaire survey, given its detail about household location.

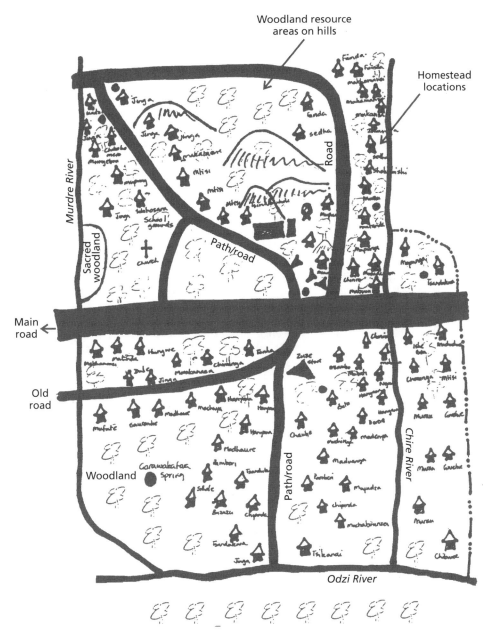

Source: Hot Springs Working Group, 1995

Figure 6.4 *Example of a resource map drawn by adults at a PRA group meeting in Jinga village, Zimbabwe. The location of each homestead in the village is shown. The major resource areas are shown (mountains and hills, woodlands next to rivers), although the area devoted to them on the map is very small (in relation to what would be shown in a map drawn to scale)*

BOX 6.2 EXPLORING THE CONFLICTS BETWEEN LOCAL COMMUNITIES AND THE ZIMBABWEAN FORESTRY COMMISSION, USING ROLE-PLAYS

Villagers were divided into groups and asked to prepare short plays depicting conflicts that have occurred in connection with their use of the village woodlands and the state forest (Gwaai Working Group, 1997). Some villagers depicted ordinary villagers in the plays, others the chief, while some took on the role of forestry officials. The major conflict identified was associated with grazing inside Mzola State Forest. While in other forest reserves people have been allowed to graze their livestock in the forest, in the case of Mzola there are complications. The communal lands are in the 'green zone', which is free from foot and mouth disease, whereas the forest area falls under the 'red zone', which is an area infested with the disease. The role-plays illustrated these conflicts. Furthermore, the role-plays and accompanying discussions on conflicts revealed the existence of several rule systems concerning woodland use. While officers of the Forestry Commission appear to bear the brunt of people's anger and abuse, some of the actions which angered people resulted from rules applied by other departments, such as the Department of National Parks and Wildlife Management or the Department of Veterinary Services.

Role-plays can thus be used to explore the different stakeholder groups who are involved in a particular issue, and to elicit some of the values held by the different groups. In the above role-plays, some of the key stakeholders were the Forestry Commission, the traditional leadership, ordinary people and farmers with large livestock herds. The key value of the forests that was held by the villagers was related to grazing. In subsequent valuation exercises, the value of grazing was determined and compared to other land uses.

---

presenting small spontaneous plays, which describe real life situations. In these plays, a situation or problem is outlined to the community group and the group takes on the roles of the people involved. The situation must be clearly outlined and must be important and relevant to the participants. Volunteers act out what a person would say or do in a given situation. Other group members carefully watch; after the play, all members discuss their reactions. The facilitator might guide the discussion by asking questions such as: What do you see in the role-play? Why do you think this happened? What were the causes of the problem? What can we do about this problem?

Role-playing can be used to convey messages on conflict over forest resource use or to encourage people to be self-sufficient. Role-plays are more useful when the process of developing them involves the local people who will act out the play. Augusto Boal, whose work has influenced theatre for development, notes that 'theatre is a weapon and it is the people who should wield it' (1979, p22). By asking people to dramatize scenarios of their everyday lives, and issues of access to forest products, this encourages them to create their own solutions. Box 6.2 illustrates how role-plays were used in Zimbabwe to investigate conflicts between local communities and the Zimbabwean Forestry Commission over access to forest products (Gwaai Working Group, 1997). Role-plays can be used to portray non-conflict situations as well.

Role-plays can also be used as a first step in a valuation exercise. For instance, they have been used to explore how forests are valued; these values are then used in subsequent scoring exercises or even in household questionnaires. In one example, participants were asked to play different key individuals in the village, and to make plays that brought out three different values of forests (Campbell et al, 1997b; see Box 6.5). Some participants played ordinary villagers, while others played craftspeople, traditional doctors or spirit mediums. While the plays were in progress, other villagers wrote down the values being expressed. These were then discussed in the full group meeting, and in this way a certain set of mutually exclusive values were identified for the subsequent ranking and scoring exercises.

Ranking and scoring of benefits to forest users

Ranking and scoring techniques have been the main way that PRA tools have been used for forest valuation. There are a number of ranking and scoring tools, including pairwise-ranking, simple ranking/scoring, wealth-ranking, matrix-ranking/scoring and contingent-ranking/scoring. Ranking and scoring are tools that encourage people to make and then evaluate choices (whether individually or in a group). They are regarded as simple and inexpensive tools for generating information about how people make choices, why people make choices and what choices they make. The following sections on 'The weaknesses of ranking and scoring methods' and 'Questionnaires, time and participation' discuss whether they are, indeed, simple and inexpensive. One of the ideas behind ranking and scoring techniques for generating valuation information is that these methods may be more appropriate in subsistence-based

economies where monetary values are less familiar.

What is the optimum size of the stakeholder groups for ranking and scoring forest benefits? Experience suggests between approximately 8 and 15 people should be manageable; it is difficult to maintain balanced and fair participation with larger groups. A second issue is whether to divide up men and women. While it is true that men and women often value products and services differently, it may be useful to have mixed groups since men know more about some benefits and women know more about others. Mixed groups can therefore often result in a more informed relative ordering where the stakeholder group is genuinely mixed.

The following sub-sections describe some of the ranking and scoring techniques that can be used for forest valuation. Since these PRA techniques are the major ones used for valuation, they are discussed again in 'The weaknesses of ranking and scoring methods'.

The pairwise-ranking technique

Pairwise-ranking is a technique in which each individual item among a group of items is compared with every other item, thereby determining a ranking of the full set of items (see Box 6.3). The tool can be used for determining the main preferences and priorities of individuals or groups for a set of items, such as forest benefits, government services and seed varieties. It can also be used for ranking problems, sources of income and the effectiveness of institutions. The priorities of different groups can be compared if the different groups undertake similar exercises (in this case, the different groups should use similar items in the pairwise comparisons).

Simple scoring and ranking

Rather than performing pairwise compar-

Box 6.3 Undertaking a pairwise-ranking exercise: a case study of local income sources in a savanna woodland area

During a pairwise-ranking exercise, one should obtain a list of items to be ranked (eg income sources). The items should be tabulated in a matrix with the items listed on both axes (see Table 6.2). Each cell in the matrix represents a paired comparison of two items. The respondents should make pairwise comparisons of all items and record the item that is preferred in the appropriate cell. At each comparison, the reason for making the selection can be ascertained from the respondents. When all the cells in the upper triangular part of the matrix are filled, the frequencies of a preferred item can be summed, and a preference rank determined. The final result should also be discussed with the participants. .

Table 6.2 is an example of a pairwise-ranking exercise with women from Gudyanga village in Chimanimani, Zimbabwe; the aim was to determine the contribution of *Adansonia digitata* (baobab) to household cash income. The pairwise-ranking showed that baobab products make a relatively high contribution to cash incomes. These products were ranked second on the list of sources of cash income. As highlighted elsewhere in this chapter (see Box 6.10 and the section on 'The weaknesses of ranking and scoring methods'), these ranking results may be difficult to interpret. They can be used to obtain overviews of situations, but more detailed survey-based approaches may be necessary if greater detail and accuracy are required.

When too many categories are being compared, the exercise takes inordinately long to complete. In addition, there is often much debate within the group on the ranking of pairs, at least at the start of the exercise. This may occur because different members of the group hold different values about the items under consideration. The discussion may in itself be important to understand differentiation among the community; but it can lengthen the exercise to the point where participants become bored. Thus, it is essential to have no more than ten categories to be compared.

Table 6.2 *Example of a pairwise-ranking exercise from Gudyanga village, Zimbabwe: a comparison of the sources of cash income*

	Brick moulding	Brachy-stegia bark craft	Baobab bark and fruits	Palm crafts	Petty trading	Sisal crafts	Livestock	Beer-brewing	Score	Rank
Crops	Crops	Crops	Crops	Crops	Crops	Crops	Crops	Crops	8	1
Beer-brewing	Bricks	Beer	Baobab	Palm	Markets	Sisal	Livestock	✗	1	8
Livestock	Livestock	Livestock	Baobab	Palm	Markets	Sisal	✗		3	7
Sisal crafts	Bricks	Sisal	Baobab	Palm	Markets	✗			3	6
Petty trading	Bricks	Markets	Baobab	Palm	✗				4	5
Palm crafts	Bricks	Palm	Baobab	✗					5	4
Baobab bark and fruits	Baobab	Baobab	✗						7	2
Brachystegia bark craft	Bricks	✗							0	9
Brick-moulding	✗								5	3

Source: Nemarundwe, unpublished data

BOX 6.4 UNDERTAKING A SIMPLE SCORING EXERCISE

Once the range of benefits has been identified, possibly with the assistance of participatory mapping of forest-product flows or role-plays, it is sensible to develop a pictorial representation of the benefits (this is vital if literacy is low). The benefits can be represented on cards or on a large piece of paper placed on the ground or table, ideally using representative materials (eg leaves to represent tree fodder). If relying on pictures, these can be drawn by the villagers in order to increase their interest. However, unless there is a local artist present this can use up scarce participant time. It is not very efficient if a group of busy villagers waits around while someone painstakingly draws the pictures (and once someone has started, it is not easy to ask them to speed up, or stop, the job). One idea is to have at least some pre-drawn cards of the expected benefit categories.

The participants are then given a fixed quantity of stones, beans or whatever appears to be an appropriate local measure, and asked to allocate them on the cards or piece of paper. This exercise is often slow to start, and one person has to make the first attempt. Gradually, the others should join in by shifting stones from one benefit type to another. Usually the interest and tempo pick up as participants get interested, and there can be frequent and even heated exchanges of stones until, with time, a consensus is reached. The facilitator obviously has an important role in preventing more dominant participants from taking over, and encouraging more reticent villagers to take part.

In such exercises it is important to ensure that what is being compared is clearly defined. For instance, in comparing wild fruits with fuelwood, is the comparison in terms of what is out in the wild (the stocks) or what is being used (the flows)? Examples of results from ranking exercises can be found in Boxes 6.5, 6.7, 6.9, 6.10 and 7.5, while the problems of ranking and scoring techniques are tackled in 'The weaknesses of ranking and scoring methods'.

---

isons of items, the full range of items is considered and the most important item is ranked first or scores highest (see Box 6.4). This tool can be used for the same kinds of comparisons as with pairwise-ranking. It is a useful way of gaining insights into the way local people value forests, and in particular the relative importance they put on tangible and non-tangible benefits.

Simple scoring and ranking can lead to discussions on why some benefits are rated more highly than others, and the technique generally assists the analyst with a better understanding of the relative importance villagers give to market and non-market benefits. Researchers should follow the discussion during the process of scoring; this can give further insights on why participants rank some benefits higher than other individuals do. An example of a simple ranking exercise is given in Box 6.5. The box illustrates that the ranking of values for forest benefits, derived during the PRA exercise, was somewhat similar to the results obtained by ranking the monetary values obtained for the benefits from a market-based questionnaire survey.

Matrix-ranking and scoring

Matrix-scoring and ranking can also help elicit the relative values of forest resources. It is usually done on a two-dimensional matrix, with the items listed on one axis and the characteristics of the items listed on the other axis (see Box 6.6). The aims of creating matrices are to discover the preferences of individuals or groups for the various characteristics of single items.

Box 6.5 Ranking benefits from savanna woodland resources in Zimbabwe

A valuation study took place in two villages of the communal areas of eastern Zimbabwe (Hot Springs Working Group, 1995). The objective was to assess woodland values in smallholder production systems, and to examine how these values might vary with resource scarcity, gender and tenure. One part of the study was the valuation of forest benefits using market-based and household survey techniques. Another part of the study was a PRA-based process of identification, ranking and, in one of the villages, scoring the benefits. This component included the more intangible benefits from forests.

In the first village, each sub-group was asked to prepare a role-play demonstrating woodland values from its perspective. From these role-plays, 19 values were recorded. The more artistic villagers then drew a picture of each benefit three times on large cards. A copy was then given to the men's, women's and boys' groups. Each group was given 100 beans to allocate between the 19 cards. A matrix-ranking and scoring exercise was then undertaken. There was a slight variation on this approach in the second village, where six groups ranked, but did not score, the benefits. From this an average overall ranking for each benefit was derived, with the lowest numbers representing the highest ranking (eg 1.3 for drinking water, 3.5 for thatch grass, 5.5 for firewood, etc).

The value of the quantified woodland products using the market-based approach came to US$50 and US$85 per household in the two villages (or US$5 to $17 per hectare per year). In the ranking exercises in both villages, indirect use or environmental service values were ranked highest, especially water-related benefits (see Figure 7.6 for the results of one of the PRA groups). An interesting part of the analysis was the comparison between the ordering of the benefits based on the market valuation and the ranking of these same benefits in the PRA exercise (see Table 6.3).

Table 6.3 *Comparison of the values of specified forest products derived from market-based and PRA approaches (the ranks from the PRA are exclusive of the other benefits that were ranked)*

Product	Jinga		Matendeudze	
	Market rank	PRA rank	Market rank	PRA rank
Firewood	1	2	1	2
Fruits	2	3	2	3
Poles	3	1	3	1
Mushrooms	–	–	4	4

With the marked exception of poles, the ordering of the extractive use values was similar in the market analysis and the PRA ranking exercise. In the second village, such 'benefits' as water retention, rain-making functions, 'inheritance' and the importance of woodlands as sacred sites were ranked higher than market-based benefits such as firewood. Where non-market values were ranked higher than the market-based values, an obvious inference is that the actual but unknown non-market values are likely to be higher than the market values recorded. The problems of PRA ranking and scoring techniques are presented in 'The weaknesses of ranking and scoring methods'.

BOX 6.6 UNDERTAKING A MATRIX-SCORING EXERCISE

Large sheets of paper and pens, counters and samples of items to be ranked are needed for this exercise. The matrix is used to compare (through scoring) the characteristics of different trees, the features of different soil and water conservation methods, the characteristics of different varieties of a crop, etc.

Participants first identify the items to be ranked and collect samples or make lists of the items. They are then asked to give the criteria they use to rank the items. The criteria or characteristics are then listed at right angles to the first list of items (see Table 6.4). These criteria are the judgements people make when evaluating and analysing the qualities or merits of items being discussed. Counters are then allocated according to preference against each characteristic or criteria. It is not just the matrix that matters, but the discussion that ensues and the knowledge shared. To be truly beneficial the process must include sensitive and perceptive interviewing.

Table 6.4 *Example of a matrix for indigenous fruit tree characteristics, with counters distributed among species and characteristics. The exercise was conducted in the Hot Springs area in Zimbabwe*

Type of fruit	Characteristic Taste	Size	Uses	Storability	
Adansonia digitata	0 0 0 0 0 0 0 0 0	0 0 0 0 0 0	0 0 0	0 0 0 0 0 0 0	25
Berchemia discolor	0 0 0	0 0 0 0 0 0	0 0 0 0 0 0 0 0 0	0 0 0 0 0	23
Azanza garckeana	0 0	0 0 0	0	0 0	16
Diospyros mespiliformis	0 0 0 0 0 0 0 0 0 0 0 0	0 0 0 0 0 0	0 0 0	0 0 0	16
	26	21	16	17	80

In this example, 80 stones were given to the participants to distribute among the cells. In other cases, a certain number of stones can be given per row or per column, depending upon the important axis of variation that the researcher wishes to explore. The results suggest that taste is the most important characteristic of the fruits, followed by size. The most desirable species is *Adansonia digitata* followed by *Berchemia discolor*. In this matrix-ranking, the right column can be used to give an idea about relative values of the different species. If the numbers of counters were held constant for each column, then the relative values for each of the species for each characteristic would be evaluated. If the numbers of counters were held constant for each row, then the focus would be on identifying, for each species, the relative importance of each of the characteristics.

In this example, one could ask what the respondents were evaluating. Were they putting a value on their use of different species (ie equivalent to $/household/annum) or were they putting a value on a unit of each species (ie equivalent to $/unit volume)? Were they thinking about availability (ie equivalent to $/hectare)? This illustrates some of the problems with scoring techniques.

Contingent-ranking and scoring

Economic analysis of community- or local-level forestry activities faces the problem of how to value the non-market benefits realized by local people (although even valuation of the more tangible benefits can often be difficult). One approach, which is an extension of the scoring approaches discussed above, has become known as contingent-ranking, and there are at least two published attempts to do this (Lynam et al, 1994; Emerton, 1996). It is also an extension of the contingent valuation method (CVM), used mainly in environmental economics, in which a hypothetical market for environmental services is created and beneficiaries are asked to state their willingness to pay (WTP) for such a benefit, or how much they are willing to accept (WTA) in compensation for loss of such a benefit. There have been several applications of CV in a forestry context, but considerable controversy surrounds the results (Davies and Richards, 1999). Chapter 4 provides a detailed description of the CVM approach.

The main difference with CVM is that, in contingent-ranking, respondents are not asked to place a monetary value on the environmental or other non-market benefit itself. Instead, a range of benefits are ranked and are then scored relative to each other, using as a basis one or two well-known (in terms of value) amenities that serve as 'anchor' values. The respondents' WTP for the anchor amenity or benefit is then used as the *numeraire* to estimate the WTP of the other non-market benefits. Thus, the essential attraction of contingent-ranking is that it allows the respondents to place values on a range of non-market items, simply by scoring the items.

Box 6.7 presents an example from Zimbabwe in which the contingent-ranking method was used to estimate the indirect use and subsistence benefits from agroforestry accruing to small farmers (Campbell et al, 1991; Lynam et al, 1994). The farmers were asked to rank and score ten benefits from multipurpose agroforestry trees, as well as a hand-pump borehole and a latrine. The farmers' WTP for the borehole served as the *numeraire* with which to value the forest products and services, according to their relative scoring. Another study from Kenya employed contingent-ranking in the exploration of subsistence and indirect use values that were threatened by external land-use pressures (Emerton, 1996). In this case, the *numeraire* was a castrated bullock as a component of the local economy representing wealth and a common medium of exchange. The problems with such techniques are highlighted in 'The weaknesses of ranking and scoring methods'.

An alternative approach to a contingent question was attempted by Adamowicz et al (1997). In a PRA group meeting with a community on the edge of Mzola Forest, the group was asked what individuals were willing to accept for compensation for the loss of access to some forest area. This is described further in Box 4.7.

The barter game for valuing non-marketed forest products

It can sometimes be very difficult to establish reliable values for home-consumed forest products that are not marketed – for example, fuelwood, wild fruits and game. One experimental exercise that still requires academic scrutiny has been used in Bolivia (Vallejos et al, 1996) and Nepal (Richards et al, 1999b). It is the 'barter exchange' game presented in Box 6.8 (see also 'Related goods approaches' in Chapter 4). This is another variation of contingent valuation; a hypothetical

---

BOX 6.7 USE OF CONTINGENT-RANKING TO VALUE FOREST BENEFITS IN ZIMBABWE

Market-based and contingent valuation (CV) techniques were used to value the benefits from multipurpose trees by farmers in one of the communal areas in Zimbabwe (Lynam et al, 1994; Campbell et al, 1991). A household questionnaire was administered to 359 farmers. The questionnaire was developed in a half-day workshop of academics and professionals, and was refined following a field test. In the first part of the questionnaire, ten cards were laid out before the respondents representing the main (previously ascertained) tree benefits. Two cards representing non-tree commodities were also included as 'anchor' values: a hand borehole and a 'Blair' latrine. The cards were ranked in order of importance by the respondent, who was then asked to score the 12 goods and services with 50 matches. Three main questions were put to the respondents:

1. What would they be prepared to pay to have the (hypothetical) opportunity of joining four other households in sinking a borehole and installing a hand pump, with success guaranteed and an interest-free loan to be paid back over five years? This was the WTP for their share of the borehole.
2. What compensation would they accept from the state if it subsequently decided to destroy the borehole? This resulted in a WTA value to be used as a validity check.
3. What would be their choice between a share of the borehole and (a specified) commodity (performed for five commodities, decreasing in value from about Z$35,000 down to Z$90)?*

Matches allocated to each category were then standardized against the points allocated to the borehole. Thus, each benefit was expressed in terms of its borehole equivalent, and then multiplied by the WTP borehole value. Validity checks were carried out on the WTP estimate and proved satisfactory (eg comparing the WTP with the cost of building a borehole). With regard to the WTP of the forest benefits, it was found that the products consumed by households had the highest values, followed by inputs to crop and animal production, and then cash, health and social service values (see Table 6.5).

Table 6.5 *Mean and median values of trees per household for various goods and services in Zimbabwean communal areas*

Good/service	Mean WTP ($Z)	Median WTP ($Z)
Fuel	373	500
Farm/house materials	290	400
Crop production	222	333
Animal feed	181	144
Ecological services	175	257
Food	136	200
Shade	102	150
Cash income	82	125
Health	71	100
Social services	46	47

When the total was converted to an annual benefit stream using discount rates of between 5 per cent and 20 per cent, the annual benefit came to a range of Z$84 to Z$336 per household. This came to 3 per cent to 50 per cent of household income (eg crop, remittances, livestock, local work, small-scale industries), depending upon the area and discount rate. While the method is intuitively appealing, there are numerous problems to be overcome (see 'The weaknesses of ranking and scoring methods').

* Values expressed in Zimbabwean dollars: Z$1=US$0.32 in 1991.

Box 6.8 Establishing values for NTFPs using the barter-exchange approach

A survey was carried out of NTFP consumption by families involved in the Lomerio project: a long-running indigenous community forest-management project in Amazonian Bolivia (Vallejos et al,1996). To complement the consumption data on fruits, firewood and game, a two-day workshop was held with a group of representative villagers in order to estimate unit values.

For valuation of the 13 most important wild fruits (according to the participants), the participants divided into two groups and developed bartering games based on 1kg bags of salt sold in the community for 0.50 Bolivianos (Bs) per kilogram.[*] The villagers and facilitators took turns to act as salt owners and fruit sellers. The villagers first expressed their WTP for the fruit in terms of kilograms of salt that they would be prepared to pay the fruit sellers.

The unit value of firewood was found in a similar way. The following variations were developed:

- Five villagers, each with 4 bags of salt, offered to buy firewood (any quantity) from a villager with 1 haz (20kg) firewood.
- Five villagers, each with 1 haz of firewood, offered to buy salt from a villager with 20 bags of salt.
- Following a lottery, six villagers each received 1 haz of firewood, while six villagers each had 6 bags of salt; they then bargained away their firewood/salt up to the point where they wanted to take back the combination of salt and firewood to their homes. In four of the six cases of villagers with salt, 1 haz of firewood was purchased for 3 bags of salt; but in one case the salt owner was not prepared to sell any of his salt for firewood.

The first variation in the firewood valuation exercise resulted in a group discussion in which it was decided that the firewood value was too high. Most transactions were in the 2 to 4 bags of salt range, and the final decision was to use 2 Bolivianos per haz, equivalent to 4 bags of salt.

[*] Values reported in Bolivianos, which at the time of the survey were US$1 = Bs 5.1.

---

market is established in which items with known values are traded with items of unknown values. The exercise provides a very useful basis for a group discussion in order to obtain a consensus view on appropriate values.

Strengths and weaknesses of PRA for understanding forest values

The potential of PRA

Numerous weaknesses in household questionnaires have been identified (see, for example, IIED 1997).[1] The perceived shortcomings of household questionnaires have resulted in an increased focus on, and use of, participatory or PRA-based research methods. Included in this shift to PRA has been the attempt to quantify and value forest benefits and, to a lesser extent, costs. The great strength of PRA, in comparison with more traditional research methods, is its ability to go beyond numbers and discuss vital qualitative issues, as well as explore, qualitatively, differentiation across forest users or other stakeholder groups. This chapter does not subscribe to the view that quantification is simple, or even desirable, with PRA tools (see 'The weaknesses of ranking and scoring methods').

One of the strengths of PRA is that it provides a range of tools and techniques that enable flexibility. In addition, it provides some visual impact and comparative analysis. PRA also encourages diverse perspectives, multiple interpretations and a multidisciplinary analysis (IIED, 1997; see also Chapter 7). In development research, PRA exercises create an opportunity for the researcher to observe power relations in the study context (Goebel, 1998). For instance, leaders and influential individuals can be identified and gender relations can be analysed.

Using PRA carefully

All work to be initiated in local communities requires that researchers develop the appropriate rapport with communities. If PRA is used as the entry point to the villages when there is no rapport between the team and the villagers, this may raise local people's suspicions. If the PRA team is unaware of the villagers' cultural background, the village's social stratification and the village's past experiences with similar interventions, the danger of rejection exists (Leurs, 1996). The PRA process usually creates expectations and if these are not met, disappointment can result in frustration and mistrust. Despite the power of PRA, when well done it is not a quick fix to complex problems – although there are some who wish this were the case.

It should be noted that PRA takes place within local power processes. PRA groups tend not to be homogeneous, and thus results can be biased towards the wealthier and more powerful – these people are often dominant in a group situation. Power relationships are often visible during group exercises. Mosse (1994) refers to PRA activities as 'public' social events. The public nature of the PRA exercises can create and exclude particular knowledge, as the output may be strongly influenced by existing social relations. The participants may project the view of the most powerful to create consensus, with the interests of the powerful being portrayed as the 'common interest'. To reduce power-related data distortions, the facilitators may compare and contrast results of PRA and household-level exercises. Observation of participation in various community activities is also important.

Just as household questionnaires can be exceptionally poorly conducted, so can PRA. Thus, many of the examples of PRA observed by the authors and their collaborators would be better described as RRA, given their rapid and extractive nature. In some cases PRA differs little from what has been described as rural development tourism. There are also too many instances where triangulation of results is not conducted, and where there is poor facilitation and lack of a critical attitude to the emerging results.

The weaknesses of ranking and scoring methods

Ranking and scoring techniques often form the cornerstone of some of the attempts to derive quantitative values for forest products. This section highlights several of their associated problems. Boxes 6.9 and 6.10 provide two further case studies that illustrate the difficulty of using ranking and scoring methods. Box 6.9 illustrates the problems of comparing different groups of participants when different facilitators are employed and benefit categories are not the same across groups. In addition, categories may not be mutually exclusive, and therefore there may be double-counting of benefits. Cavendish (personal communication) has pointed out similar problems in the attempt to rank forest benefits in Zimbabwe (Hot Springs Working Group, 1995; see Box 6.5). These include the following:

- overlapping between such user-defined benefits as the 'inheritance' value and the 'aesthetic' value of forests;
- the problem of disappearing and/or 'irrational' values (eg older people value certain trees because they 'attract rain', while younger people do not);
- a focus on benefits means that negative values are missed (eg forests may act as a hiding place for thieves and crop-damaging animals);
- the confusion between ranking flow variables versus stock variables.

Box 6.10 illustrates the unreliability of quantitative values. Different results may be obtained from similar groups, for no simple or apparent reason. Experience has shown that the more elements that need to be compared, the more spurious are the resulting data.

While experience and practice can result in more consistent benefit categories, there is a conflict between participatory benefit identification and attempts by the analyst to control categories in order to increase comparability. It was also noted that occasionally groups had to be prompted into considering the less tangible benefits, and so the results were biased by the 'leading question' problem. Ultimately, the scoring techniques are much more of an art than a science. The quantitative information is much less important than the qualitative insights that are gained during the processes and during the discussion surrounding the ranking and scoring.

Taking the scoring techniques to the point of the contingent-ranking method, and the calculation of monetary values, is even more dubious, given the serious questions that have been raised (Cavendish, Luckert, Bishop and pers comm; Clarke et al, 1996). Difficulties include the following:

- It is not valid to convert a rank to a cardinal value. While it may be possible to convert scores to cardinal values, one has to be sure that the scores that are given for items refer to clear units and scales. For instance, is the value being given to fuelwood for a bundle of wood or for yearly consumption? Is fruit value in an equivalent unit?
- Defining benefit categories that are independent and consistent: there is a tendency to mix up flow (resource use) and stock (resource availability) values, so that in some ranking and scoring techniques, flow and stock variables are inappropriately compared (see Box 6.6). Double-counting problems can result from overlapping benefit categories; and benefit categories tend to be poorly specified in terms of their quantity (area), duration and quality.
- The use of a 'durable good'-type *numeraire*, such as a cow or radio (as in the Emerton, 1996, study), which is differentially owned by the respondents (whether they own an item, and how many they own, will clearly influence their valuation of it). See 'Ranking and scoring of benefits to

BOX 6.9 THE RANKING AND SCORING OF FOREST BENEFITS IN NEPAL

During PRA with the Dumre Sanne Community Forest User Group (FUG) in Nepal (Richards et al, 1999b), wealth-ranking was undertaken (stakeholder groups can be wealth-ranked using the wealth-ranking technique – Abdi et al, 1997). A subsequent exercise ranked and scored the forest benefits for the different wealth groups (see Table 6.6).

Table 6.6 *Ranking and scoring of forest benefits from Dumre Sanne Community Forest User Group in Nepal*

Activity	Poorer group Rank	Poorer group Score (%)	Middle group men Rank	Middle group men Score (%)	Middle group women Rank	Middle group women Score (%)	Richer group Rank	Richer group Score (%)
Fuelwood	1	10	5	11	8	7	2	13
Timber	6	8	8	6	11	5	7	6
Poles	8	6	7	8				
Tree fodder							6	8
Leaf litter	8	6			8	7	7	6
Grass fodder	1	10	1	17	3	10		
Thatching grass			5	11				
Grazing	6	8			3	10	2	13
Resin (and cones)	1	10			8	7	11	4
Agricultural tools	1	10			3	10		
Nursery seedlings	8	6			1	12		
Water	1	10			1	12		
Red soil	11	4	9	6				
FUG employment			2	13				
FUG fund					7	9		
Cash income							7	6
Soil conservation/ erosion control	1	10	2	13	3	10	4	11
Community management and control							1	17
Group solidarity			2	13			5	10
Clean air							7	6

It is not advisable to treat such exercises in a more quantitative way. The main problem is that 'like is not being compared to like' and each ranking and scoring exercise has to be treated as a separate independent experience. In Table 6.6, the four stakeholder groups included different values according to how the participants understood the exercise, and probably according to the prompting of the facilitator (in this case, there were two facilitating teams). Two groups (those facilitated by the second team) tended to introduce wider and less tangible benefits of community forestry, such as 'group solidarity' and 'community management and control'. The wealthier household group mentioned cash income as a separate item while still maintaining other categories, such as resin, which is sold for cash. Thus, this is a case of double-counting. There was an arguably 'irrational' value of 'clean air' listed by the richer group. This made it very difficult to compare the results across the groups (see Figure 6.5).

Photograph by Michael Richards

Figure 6.5 *Members of the Cwebe community in Transkei, South Africa, performing a ranking and scoring exercise about forest benefits. While it was inadvisable to use the results emerging from such exercises to derive monetary values, the discussion surrounding the exercise was often ardent and insightful. Thus, the qualitative results from such exercises may be a more important output than the numbers in the matrix*

forest users' and Box 6.7 for a discussion of *numeraires*.

- The questionable validity of using a market-based *numeraire* in a primarily subsistence rural economy, and the resulting unreliability of the numbers.
- Observed farmer behaviour: for example, in some case studies firewood has been ranked higher than fruit production; but observed farmer preference is for the planting of fruit rather than firewood trees. It is likely that there may be confusion in what is being ranked, in terms of units and scales.

The conclusion here is that the pitfalls of participatory valuation attempts are serious, and any results from such exercises should be treated with great caution, especially attempts to develop monetary values from scoring exercises. In most situations, it would be better to value tangible values more accurately, and to explore the non-market values in a more qualitative fashion or more rigorously, as described in Chapter 4. Finally, while this chapter focuses on the problems of using PRA to quantify forest values, the problems of using PRA to quantify ecological resources have also been noted (Shanley, 1999; Shanley et al, 1997; Cunningham, 2000).

BOX 6.10 COMPARING THE CASH INCOME SOURCES DERIVED FROM PRA GROUPS AND HOUSEHOLD SURVEYS

Recent data from a workshop in southern Zimbabwe seem to confirm the unreliability of scoring exercises (Richards et al, 1999a). Researchers were particularly interested in determining the level of income from palm products (sap for palm wine and leaves for craftwork) in relation to other income sources. One exercise to rank and score sources of household income was repeated in two PRA sessions. Most of the participants in the second session (approximately 30 individuals) had been in the first session the day before (approximately 50 individuals). This gave results that were disturbingly different for the two sessions. Table 6.7 presents these results, and also those derived from a household survey in the same area; the results of the latter were closer to those of the first larger PRA session (except for the palm products). One observation is that PRA groups of this size are almost too big for effective participation and very difficult for the facilitator to control.

The survey suggested that palm craft sales were an important component of cash income, while very few households that sold palm wine were sampled, suggesting that palm wine is a very infrequent cash-income source (see Figures 6.6 and 6.7). The survey suggested that crafts were the third most important income source. The PRA sessions gave rather different results, with palm wine ranked as the second and sixth most important source of income in the two PRA sessions, and palm crafts as the third and seventh source.

Table 6.7 *Ranking of cash income sources by craftswomen in southern Zimbabwe: a comparison of the results from two PRA sessions and a household survey*

| | PRA | | | | Survey Craftswomen | |
| | Session 1 | | Session 2 | | | |
Income source	Score	Rank	Score	Rank	Share of total income (%)	Rank
Buy/sell various goods	11	1	6	5	1.3	9
Livestock sales	10	2	4	6	38.7	1
Crop sales	10	3	–	–	18.9	2
Beer-brewing	10	4	3	7	2.6	6
Formal employment	8	5	2	8	7.7	4
Palm wine production	7	6	8	2	–	–
Palm craft sales	5	7	8	3	18.2	3
Thatching grass	5	8	1	11	1.9	8
Casual labour	5	9	1	9	0.2	11
Brick sales	4	10	9	1	1.3	9
Nut sales	2	11	1	10	–	–
Clay pot sales	1	12	8	4	–	–
Knitting, etc	–	–	–	–	5.6	5
Other	–	–	–	–	2.6	6

Photograph by Manuel Ruiz Pérez

Figure 6.6 *Palm wine sales in Mbalmayo market in Cameroon. There was an upsurge in the market following the devaluation of the currency. The widespread sales in this area differ from other parts of Africa, where palm wine may be used more for household and ritual consumption*

Questionnaires, time and participation

A further problem of quantifying with PRA scoring exercises is that it is not possible to aggregate up the results of a PRA in the way that is possible with questionnaire survey data (see Chapter 2). In addition, multiple stratification across many different variables is not possible with PRA. Much more important in these scoring and ranking exercises is the discussion that is elicited and the qualitative insights that are gained.

With household questionnaire surveys, researchers must be aware of potential biases and must test for them. With PRA, respondent and facilitator bias can be a major problem, and there is an urgent need to make PRA more rigorous. In general, it would seem unwise to use PRA methods alone for quantification, and experience suggests that research methods should be regarded as complements to each other rather than as substitutes, as seems to have been argued in the past (Chambers, 1992). In a comparison of PRA and questionnaire surveys in southern Zimbabwe (see Box 6.11), it was shown that while some of the results may be the same, there are other areas in which results differ. Results from PRA may be valuable in order to obtain data where interhousehold variation is limited (eg labour allocation to different tasks). However, the case study in Box 6.11 indicates that attempts at quantification

Photograph by Tony Cunningham

Figure 6.7 Hyphaene coriacea *(ilala) tapped for palm wine in southern Africa*

through PRA are likely to be problematic when there is considerable interhousehold variation in production levels and/or when PRA groups are heterogeneous.

From a researcher's point of view, PRA is, apparently, very cost-effective. In the case study, the research team had at its disposal several hundred hours of more community member time than, for example, in the case of a household survey (Richards et al, 1999a). For the villagers, half-day or longer meetings are more time consuming than questionnaire survey interviews (these should last for no longer than one hour). But PRA is 'cost-effective'

to researchers only when no compensation is offered to villagers. Some form of remuneration previously negotiated with the community (eg a donation to the school, provision of medical services, etc) might encourage increased participation and greater interest by the community, not to mention improved 'efficiency' within the research team if the 'costs' of participation are more apparent. Any discussion of cost-effectiveness is also dependent on the objectives. Are the tools being used in a research context or to provide baseline data for a proposed project design, or for other community purposes? The higher cost of a participatory approach can be justified if it leads to improved participation in the project cycle.

The trade-offs between the methods in terms of research objectives and cost are inherent in the need to balance accuracy or quantification with the degree of participation. There is a continuum from more informal PRA tools and key informant discussions to more formal PRA tools (which can be quite inflexible and tedious for villagers) and sample surveys. Finally, short-cut data-collection methods, such as those discussed here, are no substitute for longitudinal research methods (including inter-alia multiple visits, household recording, participatory monitoring, physical measurement, anthropological observation, etc) in order to produce reliable analyses of economic incentives and project design purposes. However, donors tend to consider short-cut data-collection methods more 'cost-effective', especially in view of the normal time frame at the beginning of the project cycle. Such a view does not consider the high cost of poorly designed project interventions and weak participation by primary beneficiaries.

BOX 6.11 A COMPARATIVE STUDY OF PRA AND SURVEY METHODS, BASED ON AN INVESTIGATION OF THE ROLE OF ILALA PALM PRODUCTS IN RURAL LIVELIHOODS

In the Mabalauta Workshop in southern Zimbabwe, a comparison was made between PRA methods and a random sample household survey using a 'closed'-type question-naire (Richards et al, 1999a). In both exercises the aim was to estimate household income derived from ilala palm (*Hyphaene petersiana*) products. Table 6.8 summarizes a comparison of the quantitative variables recorded by the two methodologies, includ-ing any noted advantages or disadvantages of the methods used. While fairly similar results were obtained for some variables, there were also some major differences. On the basis of this data alone it is difficult to state which method is more reliable; but follow-up work of a more anthropological nature (monthly visits to a smaller sub-set of households) provides support for the household survey data.

The main area of discrepancy was in the production and sale of craft products. The PRA-based estimates were found to be unrealistically high when household budgets based on the PRA data were compared to normal household incomes in the area. Possible reasons for this were:

- Specialist producers of craft items (eg washing baskets) were more dominant in the discussions.
- There was a strategic reason of wanting to show production potential, in case a marketing project was 'in the offing'.
- There was possible confusion between production and unsold stocks.

The survey produced income data more in line with secondary data sources of incomes in rural Zimbabwe (eg as derived from Cavendish, 1997). However, it should also be mentioned that while both research exercises were carried out in some haste, the PRA exercise did suffer more in terms of not conforming to best practice. For example, there were unscheduled clashes of meetings involving the villages, one of which was with an NGO that was trying to establish a craft-marketing cooperative.

For some quantitative aspects, however, the PRA exercise produced more reliable parameters (eg the labour requirements of the various craft products). The craftswomen PRA group very carefully deliberated the time required, going through each harvesting and processing stage and reaching a consensus. The survey did not have this flexibility; enumerators also reported that respondents experienced considerable difficulty with the concept of 'hours'. While this was problematic for the PRA groups, it was possible to reach a common understanding through more extended discussion. PRA methods proved more effective in terms of differentiating and understanding the complex range of baskets and other craft products, and were also better at picking up temporal differ-entiation. There were important differences in craft and wine production in good and bad (drought) agricultural years. In difficult years, people fall back more on ilala palm products, so the products act (as do many NTFPs) as a safety net.

Perhaps surprisingly, the survey was more revealing in terms of gender differences; for example, it was easier to assess and compare returns to male and female craft producers. But the survey was not without its problems. There was some confusion, for example, about the different types of basket. Furthermore, when the (normally female) craft producer was not the (normally male) survey respondent, underestimation was more likely.

Table 6.8 *Summary of comparison between PRA and surveys in the Mabalauta Workshop*

Variable	General comment on differences and similarities	Advantages of PRA and disadvantages of survey	Advantages of survey and disadvantages of PRA
Stakeholder group as proportion of population	Similar results obtained from the two methods		Statistical representation
Ranking of cash income sources, and proportion of total household income from each source (Box 6.10)	Some similarities (eg livestock, hired labour); but PRA gave much greater share to palm products	PRA better able to pick up minor or 'niche' sources of cash income (eg revenue from CAMPFIRE); PRA exercise carried out for good and bad years; survey biased towards male cash income	Distribution of income could be assessed; problem of averaging out PRA groups with wide livelihood diversity; contradictory ranking by two craftswomen PRA groups; PRA respondents less willing to reveal remittances from illegal employment in South Africa (less anonymity?); PRA might include potential income sources
Production levels of craft products (main determinant of economic returns)	Major differences: PRA production levels very high; survey data more realistic, possible underestimates	Clearer understanding of range of craft products (some confusion about basket types in survey); survey missed temporal variation; craftsperson in household often not interviewed	Easier to identify specialist producers, who were given too much weight in PRA; PRA more prone to strategic response (hoping for a project)
Labour requirements	Higher labour inputs recorded by PRA; probable underestimation by survey	Different processing stages carefully considered and consensus reached; problems of survey: male respondents giving labour time to craftswomen; difficulty with 'hours'; missed time in dye collection	
Sale prices	Prices recorded in PRA were generally higher	Prices could be discussed, including seasonal/annual variation; mathematical derivation of prices in survey meant scope for error	Presence of foreigners in PRA might have resulted in inflated prices
End uses (% sold, consumed, barter, gifts, etc)	Similar results	PRA differentiated between good and bad years (more sold in bad agricultural year)	

To overcome some of the PRA weaknesses that may arise when assessing people's values with regard to the economic contribution of forest products to household livelihood, suggestions have been made as to how PRA can be used in conjunction with other research methods, such as economic analysis tools (Davies and Richards, 1999).

The way forward

This chapter has described the potential and limitations of participatory methods for exploring tree and livelihood values for forest-user communities. The suggestion is made that for cost-effective and reliable data collection, it is essential to combine different data-collection methods, including household surveys and PRA tools. Memory recall approaches should be backed up by observation, weighing and other methods. PRA methods may be more reliable for costs than production. Key informant discussions can be a reliable way of obtaining production and economic data but need cross-checking; due to the small number of observations, a sample survey is advisable in order to increase representation. If affordable, Wollenberg and Nawir (1998) suggest a combination of interviews, record-keeping and observation. A logical sequence would be the following (Richards et al, 1999a):

1   PRA (including, in particular, key informant discussions) should be used to gain a sound understanding of livelihood issues, and the underlying economic, social and ecological relationships, as well as to inform the design of subsequent research tools. Where detailed insights are required, more anthropological techniques may be appropriate (eg Martin, 1995). The issue of context is becoming more and more important, hence the need to understand nuances and metaphors in a given location. This points to the necessity for researchers to be familiar with the location and have an ability to speak the local language.
2   Role-plays, wealth-ranking, other PRA exercises and anthropological approaches should follow in order to define stakeholder groups and identify values.
3   Stakeholder-group PRA exercises should explore user-group objectives, trade-offs and conflicts, consider control and access to forest and other local resources, and define the limiting (or scarce) factor or resource facing that group.
4   Key informants should generate the main technical and economic parameters, complemented by, wherever possible, physical observation and time recording of key activities (Box 5.6 demonstrates an appropriate use of key informants to derive valuation data).
5   A statistically representative household survey should establish household income and production levels, as well as collect more finely tuned data on household characteristics, wealth status and representation of stakeholder groups or key informants.
6   Verification and modification of the data should be conducted by comparing data from the three sources, and taking back the survey and key informant data for discussion with PRA stakeholder groups (eg discussing key technical and economic parameters and any apparent anomalies).

Different methods could also be used for cross-checking data. Research methods from one approach can be employed to increase the confidence in another research method. For example, a sample survey may show that a few key informants are reasonably representative of the wider population. If a survey was carried out first, researchers could select key informants with the 'modal' characteristics thrown up by the survey.

A final point is that the combination of tools depends very much upon the purpose of the data collection. More participatory approaches can be more appropriate where it is felt that participation is part of a project process. Extractive data-collection systems can be detrimental when a goal is maximum participation in project design, but will be essential when quantitative rigour is required. Although PRA was not initially developed for academic research, it may have applications there. Nonetheless, caution is suggested (Adams and Megaw, 1997).

Note

1   Through careful research many of the problems can be overcome and many biases can be tested within a questionnaire survey (see Chapter 2).

Chapter 7

Searching for synthesis: integrating economic perspectives with those from other disciplines

Bev Sithole, Peter Frost and Terrence S Veeman

Introduction

At the interface of the nexus between people and plants, we have the discipline of ethnobotany. While, in the past, this discipline may have been dominated by biophysical scientists, Martin (1995) has firmly made the point that ethnobotany is a multidisciplinary subject, requiring expertise in botany, ethnopharmacology, anthropology, ecology, economics and linguistics. This book has focused on the economic issues related to valuing forests, and has made the point that valuation requires strong disciplinary skills in economics in order to avoid falling prey to numerous potential pitfalls (see 'Importance of remaining observant and critical' in Chapter 1). In Chapter 6 participatory approaches were discussed, drawing attention to the benefits of involving stakeholders in the research process

and of establishing the context within which valuation can be understood (see Figure 7.1). This chapter looks at the need for disciplinary integration (see the following section) and the difference between 'multidisciplinary' and 'interdisciplinary' ('Multidisciplinarity or interdisciplinarity?'). It then reflects on the need for incorporating stakeholders in the research process ('Incorporating stakeholders in the research agenda'). The concepts ('Some concepts that promote interdisciplinarity') and methods ('Approaches and methods to foster interdisciplinarity') that could possibly promote an interdisciplinary approach are also considered. Finally, this chapter examines the constraints to an interdisciplinary approach ('Constraints to interdisciplinarity').

The need for disciplinary integration

In understanding people and their interrelationships with forests, it is important to recognize multiple causation and multiple

objectives (Chambers, 1983; see Box 7.1). On the development side, multiple interventions are likely to be more appropriate

Note: A number of results from a PRA workshop are presented in Table 7.2.
Photograph by Tim Hoffman

Figure 7.1 *A team of ecologists and economists interacting with villagers during a PRA exercise. In this exercise, the sustainability of various resources was investigated*

than single interventions. Collaboration between economists and other scientists is needed in order to provide decision-makers and the general public with the broader assessment of the costs and benefits of different options that guide environmentally sensitive development (Ehrlich and Daily, 1993).

A broad and useful notion of science is that it constitutes systematized knowledge obtained by many different kinds of observation and experiment; that this knowledge is often context specific; and that while we may strive for certainty in our understanding, this will always remain an elusive goal (Ehrlich and Daily, 1993). A keyword in the above sentence is 'context specific'; context is defined by biophysical, economic and social variables. Work on people and their relationships to forests is inherently interdisciplinary and requires approaches and tools to help integrate information that is derived in different ways from a variety of sources, often from diverse perspectives. In the context of applied research, one should aim to offset in-built disciplinary biases and achieve a broad and balanced understanding of situations or phenomena (Chambers, 1983).

Practitioners in different disciplines can be thought of as wearing different optical lenses through which they view the complexity of interactions between people and their environment (Blaickie, 1993). Which lens one uses is heavily influenced by one's training in a given discipline. Achieving an interdisciplinary approach

199

thus requires researchers to be willing and able to try different optics. Open-mindedness and cooperation need to be encouraged, so that an interdisciplinary approach becomes a 'sixth sense' in research (Stocking, 1993).

Ordinary people and politicians are often frustrated because they do not hear simple and consistent answers (Holling, 1993). Apart from the fact that some problems do not have simple answers, there are also few relevant generic concepts, technologies and methods available to deal with complex issues (Box 7.1). Integrated approaches are therefore needed to address the often messy, but critical, issues confronting society (Hilborn and Ludwig, 1993). In reality, scientists no longer have a choice because policy-makers and donors are asking questions that require an interdisciplinary focus. Increasingly, the questions focus on issues at the interfaces of the traditional disciplines and require a wide mix of skills and experiences to address them properly.

The fast-changing needs for national action and policy formulation in natural resource management have created new imperatives and opportunities for inter-disciplinary research. Prompted by the recognition that single-discipline approaches may compound rather than mitigate problems at the human–environment interface, there has been a recent push towards interdisciplinary research and training. Yet, the complex and interdisciplinary nature of current challenges in natural resource management contrasts sharply with the way in which the relevant disciplines are organized and interact. The reality of interlocked systems and interrelated problems will not change; but the nature of research and how the disciplines are organized to address the problems can and must (WCED, 1987).

Most environmental problems require the attention of scientists in three or more disciplines, most commonly the natural sciences, sociology/anthropology and economics. Understanding deforestation, for example, requires, at the very least, data on tree biomass stocks, regeneration and growth; land use, particularly agriculture; pasture quality for livestock; household livelihoods; local organizations and institutions; and legal and policy issues. However, traditional administrative structures are not set up to facilitate inter-related work. For example, to address the deforestation issue from all angles at the University of Zimbabwe would entail drawing expertise from four faculties and six departments: Biological Sciences, Agricultural Economics, Crop Science, Public Law, Geography and Environmental Sciences and the Centre for Applied Social Sciences.

Several disciplines are routinely involved in the processes of planned rural development, although until recently sociology and anthropology were not among them. This is now changing in response to repeated failures in those development programmes that have discounted the social and human factors involved. Land-use and resource-management decisions are firmly embedded in particular socio-political contexts, in which decision-making is characterized by individual preferences and values, institutional arrangements and organizational structures that shape individual choice and action. The information derived from traditional disciplinary-based science, while necessary, is not sufficient to address this complexity: it is too reductionist and detached from policy.

Disciplines evolve through the cumulative results of 'systematically studied knowledge', which leads 'learned professionals' in, turn, to differentiate both content and concepts to the point where new disciplines are recognized (Murphree, 1997). This fragmentation of disciplines

Box 7.1 Generic features of research on complex systems

- Environmental problems are essentially systems problems where aspects of the behaviour of the systems are complex and unpredictable, and where causes, while at times simple, when finally understood, are always multiple. Therefore, interdisciplinary and integrated modes of inquiry are needed for understanding. And understanding (not complete explanation) is needed to form policies.

- The causes of problems are fundamentally non-linear in their dynamics. They demonstrate multistable states and discontinuous behaviour in both time and space. Therefore, the concepts that are useful come from non-linear dynamics and theories of complex systems. Policies that rely exclusively on social or economic adaptation to 'smoothly changing' and reversible conditions lead to reduced options, limited potential and perpetual surprise.

- Environmental problems are increasingly caused by slow changes, reflecting decade-length accumulations of human influences on air and oceans and decade-to-century transformations of landscapes. Those slow changes cause sudden changes in fast environmental variables that directly affect the health of people, the productivity of renewable resources and the vitality of societies. Therefore, analysis should focus on the interactions between slow phenomena and fast ones and monitoring should focus on long-term, slow changes in structural variables. The political window that drives quick fixes for quick solutions simply leads to more unforgiving conditions for decisions, more fragile natural systems and more dependent and distrustful citizens.

- The spatial span of connections is intensifying so that problems are now fundamentally 'cross-scale' in space as well as time. National environmental problems can now more and more frequently have their source both at home and half a world away – witness greenhouse gas accumulations, the ozone hole, AIDS and loss of biodiversity. Natural planetary processes that mediate these issues are coupled to human, economic and trade linkages that have evolved over the last half a century. Therefore, the science that is needed is not only interdisciplinary: it is also cross-scale. Yet, the very best of environmental and ecological research models have achieved their success by being either scale independent or constrained to a narrow range of scales.

- Both the ecological and social components of environmental problems have an evolutionary character. The problems are therefore not amenable to solutions based on a knowledge of small parts of the whole, or on assumptions of constancy or stability of fundamental relationships – whether ecological, economic or social. Assumptions that such constancy is the rule might give a comfortable sense of certainty but are spurious. Such assumptions produce policies and science that contribute to a pathology of rigid and unseeing institutions, increasingly stressed natural systems and public dependencies. Therefore, the focus best suited to natural science components is an evolutionary one; the focus best suited to economics and organizational theory is learning and innovation; and the focus best suited to policies is actively adaptive designs that yield understanding as much as they do product.

Source: adapted from Holling, 1993

has both positive and negative implications for scholarship. Positively, it provides the context for selectivity and the detailed concentration that is necessary, today, for scholarship. Reductionism can ensure rigour in the application of concepts and methods, and can avoid superficiality. Selectivity is necessary because our current stock of knowledge is so vast that it can obscure rather than contribute to understanding. Good scholarship has become as much a matter of knowing what to ignore as it is a matter of knowing what is necessary, of practising what might be called optimal ignorance (Murphree, 1997) (see the section on 'Features and principles' in Chapter 6).

On the negative side, the fragmentation of disciplines inhibits synthesis and reduces our understanding of context. Disciplinary superstructures consist of basic concepts, modes of inquiry, observational categories, representational techniques, standards of proof and types of explanation. As a result, individuals from various disciplines often observe, understand and communicate about phenomena in ways that are alien to one another, creating distinct and difficult barriers to interaction. In short, many conceptual, behavioural and organizational difficulties typically beset interdisciplinary research. While the focus of this chapter is on the need for an interdisciplinary approach, it recognizes that getting the balance right between unidisciplinarity and interdisciplinarity may be a greater challenge.

Multidisciplinarity or interdisciplinarity?

Depending upon the number of disciplines involved, the questions being asked and the approaches being used to answer them, research can be either unidisciplinary, multidisciplinary or interdisciplinary. Unidisciplinary research is performed by practitioners within a single discipline. In contrast, multidisciplinarity involves research into different facets of a common problem, each of which is defined along different disciplinary lines. Many environmental impact assessments, as conventionally done, tend to be multidisciplinary. For example, baobab bark is widely used for making mats for sale to tourists in south-eastern Zimbabwe, and some people link widespread bark harvesting to sooty baobab disease, a fungal disease (see Box 7.2). A suitably balanced study of sooty baobab disease would include the views of physical scientists, particularly ecologists and plant pathologists, an anthropologist to examine the management institutions and an economist to examine the role of bark utilization within the household livelihood system. If the scientists approached each of their topics from a disciplinary perspective, with minimal interactions among the researchers, this would be multidisciplinary research.

The term multidisciplinary research has often been used interchangeably with interdisciplinary research, even though they describe different types of disciplinary relationships. True interdisciplinarity involves applying a common paradigm, derived from integrating both analytical and empirical approaches, from two or more disciplines, to a single problem. For example, an interdisciplinary approach to sooty baobab disease would involve the joint focus and work of, for example, ecologists, economists and sociologists in using an integrated conceptual and empirical framework. This approach is difficult

BOX 7.2 INVESTIGATING THE IMPACTS OF HARVESTING THE BARK OF *ADANSONIA DIGITATA* (BAOBAB)

Claudia Romero and Marty Luckert

In an interdisciplinary workshop in south-eastern Zimbabwe in the Save-Odzi Valley, one of the research questions centred on livelihoods and the ecology of baobab: 'What is the role of the baobab tree in household livelihoods, and to what extent can that role be maintained?' This question was largely posed because of the widespread harvesting of bark and the potential link between bark-harvesting and the demise of baobab trees as a result of the sooty baobab disease. Baobabs are susceptible to infection by a sooty mould (Sharp, 1993). Trees with the disease may die, particularly during and soon after drought years. There appears to be a positive relationship between the degree of harvesting in an area and the degree of infestation of the baobab (Mudavhanu, 1997). If, indeed, there is such a relationship, it is not yet clear whether the high rate of infection of harvested trees is due to harvesters spreading the disease from tree to tree, or whether trees are rendered more susceptible to the disease because of bark loss.

The bark is harvested to make mats and other handicrafts that are sold to tourists. Baobab bark in the study area was harvested from trees of a wide range of sizes (25–330cm diameter at breast height) (Figure 7.2). Harvesting is reportedly more prevalent during the dry season when people have available time due to reduced agricultural activities. Bark from 97 per cent of the baobab trees in the study area had been harvested at least once. More than 50 per cent of the trees had been harvested to heights of about 2m and above (requiring local people to build harvesting platforms next to the trees). The mean proportion of trees infected by the sooty disease in the study area was 47 per cent. From the existing information, it appears that bark extraction rates exceed bark regeneration rates. Given that baobab trees are not regenerating in the study area, the sustainability of bark resource use on the long term is in severe doubt.

Although the marketing of baobab products reportedly increased during the early 1990s (Braedt and Standa-Gunda, 2000), commercialization of baobab bark products seems to have been stable during the last four years (Veeman, M et al, 2001). The drive to harvest bark has also resulted in the breaking of a variety of local rules (eg the sanctity of burial sites). In general, the local rules governing resource access have not been able to deal with the rapid rise in commercialization that has occurred during the last decade.

Livelihoods are sustained by a wide variety of activities in the study area (Luckert et al, 2001; see Figure 7.3). The baobab craft industry is an important source of rural employment in the Save-Odzi Valley, especially among women and young people. Up to 43 per cent of existing households in the study area, mostly the poorest, are involved in bark-harvesting and processing. Nevertheless, baobab bark-harvesting is not uniformly practised throughout the study area. There are villages where few people are involved, whereas in others there are numerous households involved (Luckert et al, 2001). Some villages are more endowed with the resource than others; and because of overharvesting, some harvesters have to walk to neighbouring villages to find trees from which they can collect bark.

In terms of contributing to household livelihoods, baobab activities are ranked highly, and are generally ranked second only to some kinds of agricultural production.

Numerical estimates of contribution to livelihoods bear this out; each participating person receives a cash income from baobab craftwork of approximately Z$5000 per annum (see Table 7.1). If opportunity costs of labour are subtracted, this figure decreases by about two-thirds, leaving one third of the cash income accruing as economic rent. The rent available to households seems to vary widely because there are households that are well located close to baobab trees, which greatly reduces production costs and increases the capture of economic rents.

In short, the contribution of baobab activities to rural livelihoods appears crucial. This conclusion makes the sustainability of this resource vital. Since it appears that current use rates are not sustainable, there is scope for investigations into policies and management options that could foster sustained use. The study emphasizes the value of an interdisciplinary approach to assessing the complex issues involved in natural resource management.

Table 7.1 *Quantities produced of various baobab products and gross income derived*

Product	Price (Z$)	Quantity produced per annum per producer household	Total baobab income per annum per producer household (Z$)
Bundles		68	188
Average	2.79		
SD	0.99		
n =	119		
Large mats		8	2800
Average	343.15		
SD	164.92		
n =	54		
Small mats		35	1809
Average	51.43		
SD	85.8		
n =	64		
Bags		5	136
Average	30.17		
SD	16.6		
n =	12		
Hats		7	1
Average	11.83		
SD	4.76		
n =	12		
Total per participating person			4998

SD = standard deviation.

and is, no doubt, the key reason why research involving several disciplines is usually multidisciplinary in nature rather than interdisciplinary. Cooperation among scientists, while positive and desirable, does not necessarily constitute interdisciplinarity, particularly if those in one discipline cannot influence the thinking of those in others.

Photograph by Manuel Ruiz Pérez

Figure 7.2 *Baobab* (Adansonia digitata) *trees in south-eastern Zimbabwe, showing the high degree of bark stripping that has occurred. The fibre is processed for use in making handicrafts and mats*

Incorporating stakeholders in the research agenda

Unfortunately, science has come to be regarded by many as a specialized activity somewhat isolated from the concerns and actions of ordinary people (Murphree, 1997). While governments seek to involve local people in managing the use of natural resources, they continue to base management decisions solely on scientific precepts. In this context, Getz et al (1999) argue not only for more integration among scientific disciplines but also for the recognition and inclusion of villagers as scientists in their own right. After all, these people are integral to the workings of the real world of natural resource use and management (WCED, 1987). Scientists need to get involved with managers and the public in adaptive management; without this, science is unlikely to become

the model for management.

The call, then, is to not only to strive for interdisciplinarity, but also to incorporate stakeholders in the scientific process. Participatory tools make an excellent contribution to both of these aims (see Chapter 6 and Figure 7.4). Once again, the challenge is probably to strike the right balance between unidisciplinary approaches, where great depth of scholarship can be achieved, and interdisciplinary approaches, where synthesis can be achieved. In addition, it is important to identify the appropriate approaches and tools for ensuring stakeholder involvement, while at the same time being free to use the methods and concepts of science, many of which may be intractable to all stakeholders.

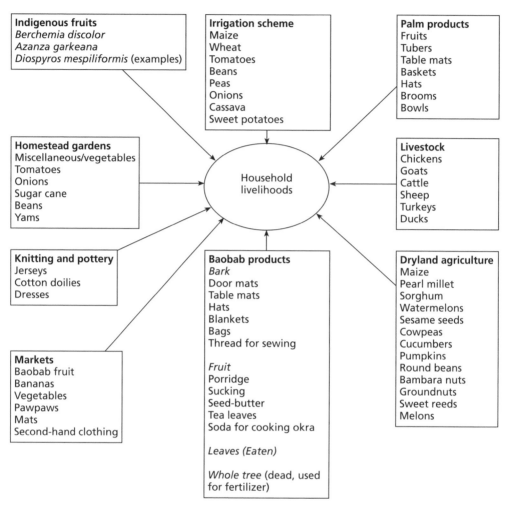

Source: Luckert et al, 2001

Figure 7.3 *Livelihoods' linkage diagram drawn by a woman's group in Tonhorai village, south-eastern Zimbabwe, showing the resources and activities making up livelihoods*

Some concepts that promote interdisciplinarity

By embracing certain concepts of natural resource management we are drawn into interdisciplinary approaches. The following concepts are discussed in the proceeding sections: sustainability, value, institutions and sustainable livelihoods.

Sustainability

The concept of sustainable development provides a framework for integrating environment and development strategies. WCED (1987) suggest that the ability to choose policy paths that are sustainable

Photograph by Tim Hoffman

Figure 7.4 *A large matrix on the ground constructed by the participants of a PRA exercise. In this scoring exercise in Hot Springs, Zimbabwe, the quantities of different kinds of resources taken from different landscape units were estimated. The group work provided an excellent opportunity for a variety of stakeholders to interact, including members of the village leadership, women and men from the village, government officials, local teachers and researchers. However, careful facilitation was required*

requires that the ecological dimensions of policy are considered at the same time as the economic, trade, sectoral and socio-political dimensions. The pursuit of sustainable development necessitates a new orientation in disciplinary relations. Sustainable development, not readily defined because it is multifaceted, involves at least four important sub-components (Veeman, 1989; UNDP, 1995) that cut across several disciplines. The growth component, initially the main concern of development economists, involves the creation of productive capacity and, as a consequence, the accumulation of capital in its various forms. The distribution component comprises issues of current-day equity among stakeholders (of concern particularly to social scientists) and involves intergenerational equity, such as what asset base to leave to the next generation, a question of interest to ecologists, economists and many others. The environmental component involves the implications of resource scarcities, resource degradation and pollution, of interest to both biophysical and social scientists. Finally, the institutional components of sustainable development embrace the participation and empowerment of people, the organizations and stakeholders involved, and the rules and regulations governing resource use. The institutional aspects are of especial interest to sociolo-

---

BOX 7.3 SUSTAINABLE FOREST MANAGEMENT, AND CRITERIA AND INDICATORS

Sustainable forest management is the set of objectives, activities and outcomes consistent with maintaining or improving the forest's ecological integrity and contributing to people's well-being, both now and in the future (Prabhu et al, 1999). Accepting such a definition implies an interdisciplinary perspective. This can be seen in the criteria and indicators (C&I) that have been proposed to monitor the status of forest management units. The criteria are drawn from numerous disciplinary perspectives, and any team undertaking work on C&I would normally be interdisciplinary. The criteria fall under six principles, namely:

1 Policy, planning and institutional frameworks should be conducive to sustainable forest management.
2 Ecosystem integrity should be maintained.
3 Forest management should maintain or enhance fair intergenerational access to resources and economic benefits.
4 Concerned stakeholders should have acknowledged rights and means to manage forests cooperatively and equitably.
5 The health of forest actors, cultures and the forest should be acceptable to all stakeholders.
6 The yield and quality of forest goods and services should be sustainable.

---

gists, lawyers and political scientists. Ideally, then, the analysis and implementation of sustainable development requires input from many disciplines and is a positive force for greater integration across disciplines.

Definitions of sustainability can vary across disciplines (Brown and Manfredo, 1987). The social perspective is concerned with human beings and their relationships among each other. Ecological definitions focus on natural biological processes and the continued productivity and functioning of ecosystems. The economic definition of sustainability may focus unduly on growth and the creation of productive capacity, with too little attention directed to the distributional and environmental consequences of growth. Even within a discipline, sustainability can have different perspectives, as, for example, the difference between sustainable yield of timber versus sustainable forest management (see Box 7.3). The latter concept is more integrative

and includes a focus on NTFPs and people. To achieve integration of the various forms of sustainability requires interdisciplinary perspectives.

Value

The valuation of natural and environmental resources provides another mechanism for greater disciplinary integration. For more than two centuries, since the days of the early moral philosophers, people have struggled with the concept of value. More recently, environmental philosophers, sociologists and social psychologists have debated on those qualities and/or attributes that give 'value' to a commodity or a resource.

We should distinguish between held and assigned values (Adamowicz et al, 1997). *Held values* are ideals, precepts and concepts about phenomena that individuals hold or groups share (Brown and Manfredo, 1987). They are associated

BOX 7.4 UNDERSTANDING THE ROLE OF TREES IN RURAL LIVELIHOODS

Allison Goebel

For the last decade, a group of scientists from varying disciplines has been investigating the relationships between livelihood strategies and woodland systems in Zimbabwe (Campbell et al, 1991; Clarke et al, 1996; Frost, 1996; Goebel et al, 2000; Grundy et al, 1993; Mandondo, 2001; Mukamuri, 1995). The work has created an intellectual field in which economic, cultural, social, historical and ecological factors in woodlands are seen as profoundly interrelated. This broad perspective has led to contextually informed economic modelling; a sharpening in the social science approach to the sense of 'value'; and ecology read through social context and economics.

Fundamental to this work was having the privilege of rural people's perceptions, including their definitions of what constitutes 'value'. Linked to this was a commitment to participatory methodologies, with many studies using PRA (Goebel, 1998; Gwaai Working Group, 1997; Hot Springs Working Group, 1995; Campbell et al, 1997b). In addition, a large component of the project involved the testing and design of economic methods, such as contingent valuation methods, observed behaviour and derived-demand techniques, within the socio-economic and cultural context of Zimbabwe (Adamowicz et al, 1997; Lynam et al, 1994).

One set of studies that was conducted in Zimbabwe relates to assessing the sustainability of people's use and management of woodlands and woodland resources. Here, the attempt has been to move beyond the simplistic notion that population increase leads to deforestation, to a more detailed understanding of how people actually use and manage woodlands in different rural contexts, and with what environmental effects. As a starting point, the assumption that peasant practices are the only component in environmental degradation was questioned. The researchers considered the broader contextual factors, particularly:

- the historical imbalance in land distribution in Zimbabwe, which has resulted in the overpopulation and degradation of African areas; and
- the savanna climate, which is characterized by recurrent drought and affects peoples' requirements and opportunities (Campbell et al, 1997a).

In addition, peasant practices and knowledge were viewed as a potential source of solutions to environmental stress, not only as a problem in need of expert scientific 'fixing'.

Another set of studies relates to resource values. Woodlands in the savanna regions of southern Africa, such as those found in Zimbabwe, are central to the livelihood systems of rural households and provide key inputs to urban households (Clarke et al, 1996). Many aspects of woodland use are increasingly being commercialized (Brigham et al, 1996); thus, understanding trends within markets, and the contribution that income from the sale of wood and non-wood products makes to households, are key both to woodland valuation and to tracing changes in household livelihoods. While some woodland products are commercialized and hence have more easily quantifiable values, most woodland products have non-market values, including use values (subsistence products), and extra-market or non-use values based on ecological, spiritual or aesthetic benefits. In addition, values vary across several lines of differentiation and can change radically in response to macro-economic events. Given this complexity, it is crucial to acknowledge that there is no one way, economically, socially or ecologically, to capture the full range of values of the woodlands to households.

The many findings from these studies emphasize several important points when considering options for sustainable woodland management and the policies that are needed to support these. The first is the importance of interdisciplinary and participatory approaches. By developing less discipline-specific means of communication, in order to ensure understanding and exchange of ideas among experts from different disciplines, project reports have also become more accessible to policy-makers. As a result, ideas coming from projects are increasingly being considered in the policy arena. Economics has provided methods by which results from different disciplines have been integrated. When informed by insights derived from socio-cultural–historical analyses, using participatory methods that capture the perspectives and values of local people, these economic approaches provide a means of valuing woodlands in a language understood by policy-makers. For example, preliminary results from the projects were instrumental in guiding the debate during the World Bank Forest Sector Review (Bojo, 1993; Bradley and Dewees, 1993).

The second main finding is the comparatively low value accorded to woodlands by local people, compared to arable land, despite the high values given to ecological services, cultural importance and the more abstract contribution of carbon sequestration (Gwaai Working Group, 1997; Campbell et al, 2000a; Kundhlande et al, 2000). Notwithstanding this, woodlands remain central to the livelihood strategies of rural households in Zimbabwe, even though their contribution to peoples' welfare is highly dynamic and often dependent upon prevailing macro-economic circumstances (Campbell et al, 1997a).

The third main finding relates to the weakness of local institutions. As a result, concern has been raised about the viability of the current enthusiasm for localizing authority for woodland management. Overemphasizing the strength of traditional authority in rural areas could spell disaster for the otherwise sound idea of devolving responsibility for management to local communities. While traditional leadership is still powerful in some areas, it is not so in others. The presence of alternative authorities (such as village development committees), factionalism and increasing numbers of young people who are more mobile and independent (and who show less allegiance to traditional norms and values as a result) all potentially undermine the traditional leadership. Understanding the conflicts among local organizations over access to resources, including those occurring across administrative boundaries, is essential when developing local management plans.

Finally, local people have been found to have a clear sense of the ecological services provided by woodlands. They also respond to resource scarcity by adopting a conservative approach to resource use, as evidenced by reductions in fuelwood consumption. This challenges some of the conventional wisdom that people's use and perceptions of woodlands are a barrier to sustainable use, rather than the building blocks for the future.

with ideas, behaviours, outcomes and experiences, and usually reflect the goals one seeks in life or the means by which one lives one's life. Groups can be differenti-ated by their reported held values, even with the existence of differences among individuals in that group (Rokeach, 1973). While held values may be adapted over

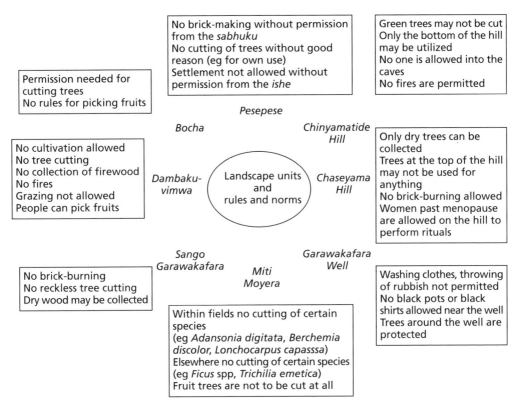

No brick-making without permission from the *sabhuku*
No cutting of trees without good reason (eg for own use)
Settlement not allowed without permission from the *ishe*

Green trees may not be cut
Only the bottom of the hill may be utilized
No one is allowed into the caves
No fires are permitted

Permission needed for cutting trees
No rules for picking fruits

Pesepese

Bocha   *Chinyamatide Hill*

No cultivation allowed
No tree cutting
No collection of firewood
No fires
Grazing not allowed
People can pick fruits

Dambaku-vimwa   Landscape units and rules and norms   *Chaseyama Hill*

Only dry trees can be collected
Trees at the top of the hill may not be used for anything
No brick-burning allowed
Women past menopause are allowed on the hill to perform rituals

Sango Garawakafara   *Miti Moyera*   *Garawakafara Well*

No brick-burning
No reckless tree cutting
Dry wood may be collected

Within fields no cutting of certain species
(eg *Adansonia digitata, Berchemia discolor, Lonchocarpus capasssa*)
Elsewhere no cutting of certain species
(eg *Ficus* spp, *Trichilia emetica*)
Fruit trees are not to be cut at all

Washing clothes, throwing of rubbish not permitted
No black pots or black shirts allowed near the well
Trees around the well are protected

Source: Hot Springs Working Group, 2001

Figure 7.5 *Institutions governing resource use in Jinga village, Zimbabwe. Different components of the landscape were derived through a PRA resource-mapping exercise. The villagers then went through each landscape unit, identifying the rules and regulations in place. While these are the stated institutions, the adherence levels were not investigated in detail; enforcement is stricter for some rules and landscape units than others*

time, in general they are seen to be stable rather than easily modified. Conflict between cultures over natural resource allocations may be due to the fact that particular cultures hold different values for natural resources.

Assigned values are defined as the relative value or worth of things (Brown and Manfredo, 1987). Assigned values tend to be associated with goods, services and opportunities. The basic values we hold (held values) determine the value we assign to goods, services and opportunities (assigned values). Assigned values are not

assumed to be stable; rather, they reflect the conditions of the goods or services, or market supply or demand, or the larger environment (Adamowicz et al, 1998).

Thus, the valuation approaches that are considered in this book are determined by held values – typically, the field of study of sociologists and anthropologists. Valuation is complicated because of the scope of values (see Table 1.1), ranging from rather intractable existence values to more direct values of traded products. The valuation of dimensions of the miombo woodlands in Zimbabwe, for example,

211

BOX 7.5 VALUING WOODLAND USE AT HOT SPRINGS, ZIMBABWE:
INTERDISCIPLINARY PERSPECTIVES

Bruce Campbell

The aim of this case study was to undertake a local-level resource valuation exercise and to assess the applicability of PRA techniques for the valuation (Hot Springs Working Group, 1995; Campbell et al, 1997b). The work was conducted in a workshop setting, involving participants who were drawn from many disciplines (ecology, forestry, agriculture, economics, sociology and anthropology). The biophysical context of the area was ascertained through ecological surveys, often done in conjunction with key informants from villagers, and PRA group techniques. Using one of these techniques, a trend towards deforestation was highlighted (see Table 7.2). Using matrix-ranking and stones to indicate quantities, the prognosis for most resources was very pessimistic. The situation was only stable for the baobab tree (*Adansonia digitata*) and *Acacia* spp (the former is conserved through traditional practices, while the latter spreads in disturbed areas). Thus, any valuation using present-day use patterns had to take into account that use may not be sustainable (see 'When own-reported values cannot be used: assumptions, omissions and other techniques' in Chapter 2 and 'Green'-accounting assessments' in Chapter 5).

Table 7.2 *Ecological sustainability analysis using matrix-ranking during a PRA workshop in Jinga Village, Zimbabwe. Participants allocated stones for different resources during different time periods to indicate the trends they perceived and projected*

Products/species	Numbers of stones allocated		
	1980	1993	circa 2006
Fruits			
Berchemia discolor	7	6	4
Adansonia digitata	12	12	12
Diospyros mespiliformis	8	6	4
Strychnos madagascariensis	7	6	3
Mushrooms	7	0	0
Honey	8	1	1
Mopane worms	2	0	0
Medicinal plants	9	6	5
Firewood/construction wood			
Colophospermum mopane	8	5	3
Combretum apiculatum	7	4	3
Brush fencing: Acacia spp	10	10	10
Reeds for crafts	8	4	3

Note: The time periods relate to Zimbabwean independence (1980), the time of the workshop (1993) and 'when the small children become adults' (circa 2006).

The workshop participants also examined the institutional arrangements that govern resource use. Using PRA tenure diagrams, overlaid on the resource units in the landscape (see Figure 7.5), there was an array of local rules and regulations. While these were adhered to in Jinga village, particularly for sacred sites, they were not adhered to in Matendeudze village. There was much intergroup conflict in the latter village, largely due to the in-migration of persons retrenched from forestry operations in Chimanimani.

Valuation was conducted using a variety of techniques. Household surveys of use patterns were conducted for mushrooms, quelea birds, firewood and construction wood. Resource values were determined using market values of these commodities and labour costs for extraction. Wild fruit values were derived from ecological surveys of the production potential of the trees in the landscape, with assumptions reflecting the labour costs and the percentage of fruits that are consumed by people (ranging from 35 per cent to 90 per cent for different species; data from key informants). More diffi-cult-to-value components of the woodland were tackled through PRA. Firstly, role-plays were facilitated to elicit the spectrum of values associated with woodlands. These values were then used in a scoring exercise, where participants distributed stones among the categories (see Figure 7.6). While the numbers derived from such scoring techniques should be treated with extreme caution, they provide an indication that in these woodland systems, people may hold ecological and sacred values very highly.*

* Scoring techniques may be intuitively appealing, and apparently simple to conduct, but they are bedevilled with numerous problems (see 'The weaknesses of ranking and scoring methods' in Chapter 6).

involves not merely timber products but also consideration of non-timber products, shade, sacred sites, grazing areas, soil maintenance, watershed functions and aesthetics (see Box 7.4). The Hot Springs Working Group (1995) highlighted the multiple values associated with the woodlands in villages in the Chimanimani District. While market and ecological values were important, religious values were also crucial (see Box 7.5).

Valuation is further complicated because natural resources are multi-attribute and thus have quantity, quality, time and space dimensions. The impor-tance of incorporating ecologists within valuation exercises is illustrated in Box 7.6 (see also Box 7.2).

Putting values on resources is further complicated by the fact that there are highly differentiated views between differ-ent age, wealth or ethnic groups regarding the significance of values attached to specific resources. The valuation of natural resources, then, encompasses many types of values and should ideally involve the interaction of many disciplines to explore and estimate the more difficult-to-measure aspects.

Institutions

To most biologists and laymen, 'institu-tions' are formal organizations such as universities, government departments, schools or environmental groups. North (1990), an economist, differentiates between institutions and organizations. Organizations, in his terminology, are the players of the game, while institutions are the rules of the game (eg rules, both formal and informal, mores, norms and govern-ment legislation). Players include universities, government departments, schools and environmental organizations, but also village development committees, the traditional village court, and house-holds. In socio-anthropological literature, institutions are regulated practices of behaviour that may be formal or informal.

While the study of institutions has been key to many sociologists and econo-mists, biophysical scientists are realizing, increasingly, that it is essential to incorpo-

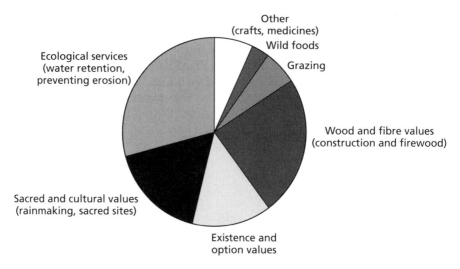

Figure 7.6 *Scores derived for different values of woodlands by a women's group in Jinga village*

rate institutional dimensions in any study of natural resource management. Institutions such as property rights (the structure of rights to resources and rules under which those rights are exercised) are mechanisms that people use to control their environment and that influence their behaviour towards one another. Institutions should be viewed in dynamic terms as products of social and political practices, and as sites where production, authority and obligation are contested and negotiated. Thus institutions will form the context within which a specified set of values can be ascertained, but the values themselves will comprise one of the drivers of institutional change (see Figure 7.5).

Sustainable livelihoods

The concept of sustainable livelihoods is an integrating concept, incorporating social, economic and ecological dimensions (Chambers and Conway, 1992; WCED, 1987). Carney (1998) and Bebbington (1999) have conceptualized 'livelihoods' as consisting of five 'capitals'

(see Figure 7.8). This conception makes explicit the interdisciplinary nature of 'livelihoods'. Scoones (1998) argues that the concept is a composite of many ideas and interests, the coming together of different strands of disciplines. Chambers and Conway (1992) regard sustainable livelihoods as comprising:

- the capabilities of individuals and livelihoods;
- the assets (including both material and social resources) that can be used by those individuals and households; and
- the resulting activities that are carried out.

A livelihood is sustainable when it can cope with, and recover from, stresses and shocks, and can maintain or enhance its capabilities and assets both now and in the future while not undermining the resource base. Murphree (1998) states that focusing on sustainable livelihoods in natural resource management reorients the emphasis on people rather than just on biodiversity conservation.

BOX 7.6 THE NEED TO INCORPORATE ECOLOGICAL DATA IN VALUATION

Bruce Campbell

Forest valuation that is not informed by detailed ecological studies can produce results that are meaningless beyond the particular time and place when, and where, the valuation was conducted. This is illustrated with results from various tropical regions.

Many valuation studies have been based on a single year of data collection. Such results can be misleading because ecological production can vary dramatically from year to year. This is especially the case with fruit from the forest, as tropical trees have particularly complex fruiting behaviour, with some years having very little fruit production and other years having considerable production. Ganesh and Davidar (1999) have shown, for a site in India, that total per hectare fruit pulp production can vary by two times between years (see Table 7.3). The variability of particular species can be even greater, and if these are important fruit species, then the value derived from the forest can change drastically from one year to the next. Shanley (1999) has shown how the production of three Amazonian fruit species changes over the course of a five-year period. Data for a ten-year period for *Vanguaria infausta*, an important fruit tree in the understorey of some types of miombo woodland, show the unpredictable nature of fruiting, with major differences among years and among the three sites near Harare (Campbell, unpublished data).

Table 7.3 *Edible fruit biomass in Kakachi rainforest, south-western Ghats, India, during 1991–1993*

	1991	1992	1993
Fruit biomass (kg/ha^{-1})	173	72	151

Note: Fruit biomass = the fruit pulp, which is edible to mammals.
Source: Ganesh and Davidar, 1999

Animal biomass can also vary from year to year, so any valuations performed in a particular year may bear no relationship to those conducted in other years. At Hot Springs in Zimbabwe, quelea birds were recorded as being a high value item in 1995; the birds are caught in their hundreds, often by boys, and sold after roasting. Subsequent visits to the area during other years showed that the birds were never as frequent as in the year of study, indicating the problems of extrapolating from a single year. The birds apparently show complex cycles of abundance and scarcity, in relation to the previous season's rainfall and the correlated crop harvest (the birds are seed-eaters). Cattle in semi-arid areas also show complex patterns of abundance, related to rainfall (see Figure 7.7); thus any attempts to value the graze and browse of woodlands in these areas have to take a long-term perspective.

Finally, much of the ecological data required to ground valuations in reality has to be collected using time-consuming ecological methods. It has been shown that production data collected using PRA can be highly unreliable (Shanley, 1999; Shanley et al, 1997; see also Richards et al, 1999a).

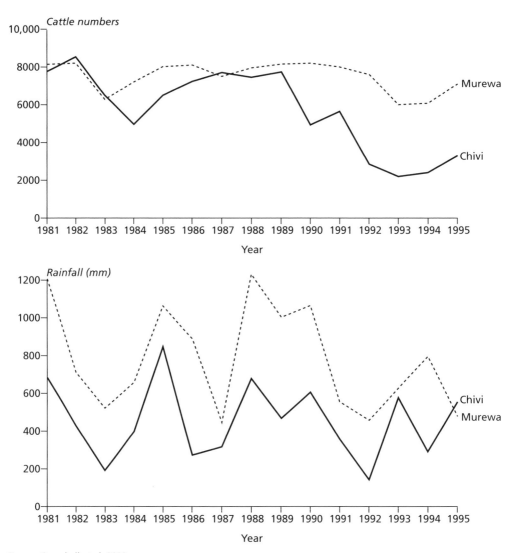

Source: Campbell et al, 2000a

Figure 7.7 *The cattle numbers and rainfall in two study sites in central and southern Zimbabwe for a 15-year period. In the slightly less arid area of Murewa, the cattle numbers do not show as marked a variation*

Approaches and methods to foster interdisciplinarity

There are particular approaches and methods that promote interdisciplinary teamwork: some relate to the process of doing the work, while others are specific methods. The following sections examine how interdisciplinary work can be initiated and how partnerships and teamwork can be developed. Participatory rural

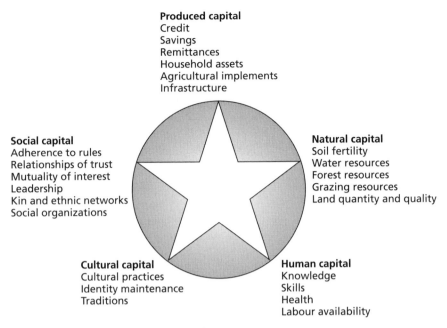

Produced capital
Credit
Savings
Remittances
Household assets
Agricultural implements
Infrastructure

Social capital
Adherence to rules
Relationships of trust
Mutuality of interest
Leadership
Kin and ethnic networks
Social organizations

Natural capital
Soil fertility
Water resources
Forest resources
Grazing resources
Land quantity and quality

Cultural capital
Cultural practices
Identity maintenance
Traditions

Human capital
Knowledge
Skills
Health
Labour availability

Source: adapted from Bebbington 1999 and Cain et al, 1999

Figure 7.8 *The capitals perspective of livelihood*

appraisal is then described as a technique for getting scientists to work together. Finally, two hi-tech tools are examined that promote interdisciplinary work: systems analysis and geographical information systems (GIS).

Initiating the work

At the start of a research programme, one of the key activities is to phrase research questions. If policy-relevant issues are to be comprehensively addressed, questions should be based on an interdisciplinary approach. It is unlikely that interdisciplinarity will occur when questions have been phrased by individuals from a particular discipline who then expect support from colleagues in other disciplines. An example of a question that requires an interdisciplinary approach is: 'What is the role of the baobab tree in household livelihoods, and to what extent can this role be

maintained?' (see Box 7.2). To answer such a question obviously requires an economist. It also requires an ecologist to understand the impacts of bark-harvesting on the tree population and whether such impacts can be maintained in the future. It requires someone with an institutional understanding who will unravel the current rules about resource use and to examine how these are changing in relation to resource scarcity.

Partnerships, interdisciplinary research workshops and teamwork

One way of bridging the gap between various disciplines is to bring scientists from different disciplines together in groups in order to work on related problems in the same geographical region. An example is the ongoing collaboration between the Institute of Environmental Studies at the University of Zimbabwe and the Department of Rural Economy at the University of Alberta. Over the last five years, both establishments have conducted five two-week workshops, each tackling a small research question requiring numerous disciplines. Each workshop involves approximately 25 to 30 individuals and is based in the field. The participants formulate the problems, select and develop methods, collect and analyse the data, and write up the work (Adamowicz et al, 1997; Campbell et al, 1997b; 2000b; Hot Springs Working Group, 1995; Gwaai Working Group, 1997; Mabalauta Working Group, 2000; Hot Springs Working Group, 2001; see also Boxes 4.7, 5.3, 6.5, 6.10, 7.2, 7.5, 7.7 and 7.8 in this book for some of the results from the workshops).

Having expertise in a range of disciplines is not sufficient to ensure an interdisciplinary approach to specific environmental problems. The framework in which interdisciplinary work is organized has to be considered. To maintain a flexible, constantly evolving team structure, considerable attention needs to be paid to the principles of project planning and monitoring, and to team-building. There is often the need to use people in the research process who have skills in facilitation. Particular attention has to be paid to ensuring good communication, developing shared vocabularies and building up mutual respect.

Participatory rural appraisal (PRA)

There is growing acceptance among many researchers that PRA is a unifying approach to conducting research. Methods used in PRA are described in detail in Chapter 6 and elsewhere (Hot Springs Working Group, 1995; Chambers, 1983; Scoones et al, 1996). PRA includes a toolbox of research methods that can be used by many disciplines, from the social to the physical sciences (Chambers, 1983;

1993; Scoones and Thompson, 1994). Teams of researchers from different disciplines can quite easily conduct PRA. Within natural resource management, PRA has concentrated on social, economic and cultural aspects, while the appraisal of physical factors has lagged behind. However, ecologists working in PRA teams have, for example, used the technique to explore woodland cover changes and have set the context for understanding valuation results (Campbell et al, 1997b; see Box 7.5). Numbers derived through qualitative assessments may be no match for detailed quantitative analysis (Cunningham, 2000; Shanley, 1999). Shanley has shown in a community in eastern Amazonia that there can be a tenfold overestimation by community members of the numbers of fruit trees occurring within their community forest.

Modelling

While simulation modelling would not be included in a book such as this only a decade ago, advances in computer technology have meant that analytical tools, such as GIS and simulation modelling, have become accessible to researchers who are not experts in these fields. The rapid uptake of newcomers to simulation modelling is illustrated in Box 7.7.

The diversity of studies carried out on the relationship between people and plants provides detailed insights into the complexity of the social, economic, political and ecological processes involved in these temporally and spatially heterogeneous systems, as illustrated by results emerging from Zimbabwe (see Box 7.8). One challenge is to synthesize these results in a coherent and meaningful way. Modelling is one option in addressing this complexity and in promoting an interdisciplinary approach. Modelling provides a common framework for integrating data,

---

BOX 7.7 ACQUIRING SKILLS IN SYSTEMS MODELLING

For a decade, a group of scientists from varying disciplines has been investigating the relationships between livelihood strategies and woodland systems in Zimbabwe. The diverse studies have provided insights into the social, economic, political and ecological processes involved in these temporally and spatially heterogeneous systems (Clarke et al, 1996; Goebel et al, 2000). One challenge is to bring the diverse results together into a common analytical framework. In order to explore the role of simulation modelling in meeting this challenge, a ten-day workshop was held. Robert Costanza and Marjan van den Belt, from the University of Maryland, facilitated the workshop, introducing participants to systems modelling. There are various graphical programming languages available that are specifically designed to facilitate the modelling of non-linear, dynamic systems. Among the most versatile of these languages is STELLA II (High Performance Systems, 1993).

Only a few of the participants had previous experience with modelling, but by the end of the workshop most had acquired the sufficient skills to be able to build their own models. At the workshop, participants worked in groups on different issues, and an integrating group built a model of a larger system using components of the models from the various groups (see Figure 7.9). The relative ease of the modelling framework (and the skills of the facilitators) is demonstrated by two facts. Firstly, the models built by the learner modellers have now been published as six papers in a special issue of an international journal, illustrating the fact that learners can rapidly acquire systems modelling skills and use them to produce research papers (Campbell et al, 2000a). Secondly, such is the relative ease of the modelling framework that learners at the first workshop have passed on their skills to other collaborators and students, who are as distant as the US and Indonesia.

---

knowledge and understanding derived from different disciplines (Gurney and Nisbet, 1998).

Modelling can take many forms, from purely conceptual 'box-and-arrow' models to simulation models. The former aims at summarizing our understanding of how different components and processes of a system are linked; simulation models are designed to mimic a real situation for the purpose of exploring system dynamics and predicting likely future outcomes under different scenarios of change. By providing a framework in which to organize and integrate existing knowledge, models can be used to assess the limits to current understanding, the adequacy of existing data, and where significant gaps in knowledge lie, key assumptions and their implications and priority areas for research. As an interactive and iterative activity, modelling can also serve to bring together researchers from different disciplines in order to identify links between them, clarify problems and identify, understand and communicate their ideas.

What model approach to adopt, and what components and processes to include, depends ultimately upon the objectives for the model. To date, most ecological models have focused on biophysical functioning; much less attention has been given to the potential use of models to integrate socio-economic and biophysical components and processes. A recent modelling initiative in Zimbabwe (Campbell et al, 2000b), aimed at developing an integrated socio-economic and ecological model of resource use in communal lands and neighbouring state

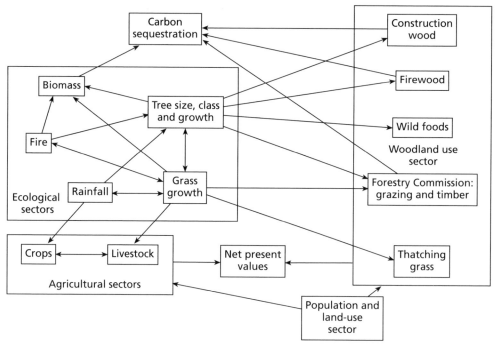

Source: Campbell et al, 2000b

Figure 7.9 *The overview of the integrated model built at the workshop of a land-use and forestry system for western Zimbabwe*

forests, highlighted several key issues. Firstly, it demonstrated the complexity of the real situation in terms of the many functional linkages among the different land-use sectors, and between ecology and economics. Secondly, it emphasized the value of modelling as a tool for exploring the implications, for local resource user groups, of different management decisions. Thirdly, it revealed significant gaps in both our ecological and economic knowledge and understanding, particularly in the paucity of data on the details of resource use and dynamics. In this way it established a future research agenda. Finally, despite the lack of data in several key areas (which meant that some stages in the modelling process involved no more than educated guessing), it was possible to develop a model that broadly depicted realities on the ground (see Box 7.8).

This initiative also revealed four major weaknesses in the way that the model was constructed. Households were treated as undifferentiated units, despite considerable evidence to the contrary (eg Cavendish, 1977; see the section on 'Analysing differences among households' in Chapter 2 and 'Influences of social differentiation within households on values and behaviour' in Chapter 8). To be useful in this regard, the model needs to capture and reflect this differentiation. This elaboration would have enabled an investigation on how the structure and wealth of a household influence decisions on resource use by its members, and how government policies impact different kinds of households. Moreover, economic behaviour as represented in the integrated model was static. A more dynamic approach is needed, one in which house-

---

BOX 7.8 CONVERTING FOREST TO CROPLAND OR EXPELLING SQUATTERS FROM FOREST? THE CASE OF MZOLA STATE FOREST

Simulation modelling has been used to integrate ecological and economic issues for a state forest in western Zimbabwe (see Box 7.7). The simulation model has five ecological sectors; five sectors covering woodland use by local people and the state forestry organization; two sectors covering agriculture; one sector for population growth and land use; a sector to cover carbon sequestration; and a sector to calculate net present values of the various uses. The model allows for the investigation of complex interactions. Thus, for example, grazing regimes can change fire regimes, which in turn affect recruitment of forest trees.

In such areas, the state has usually attempted to keep people and their livestock out of the forest. From the model it was estimated that crop production gives higher returns to land than the commercial values derived by the Zimbabwean Forestry Commission from grazing leases and timber concessions, and higher returns than the forest-use values derived by local people (see Figure 7.10). Two items are missing in this comparison. Firstly, it is unclear what proportion of the cropping value is due to government subsidies, in the form of pricing structures for inputs and outputs and drought relief. It is clear that large-scale commercial farmers would never use Mzola State Forest for crop production, given the nutrient-poor soils and extremely unreliable rainfall. The state provides drought relief very often in these regions in the form of food aid, and fertilizer and seed packs (Frost and Mandondo, 1999). Secondly, the service functions of the forest are not incorporated. Some of these functions include cultural values, modifying the hydrological cycle and carbon sequestration. However, there is almost no quantitative data on any of these services and therefore it is difficult to evaluate them. Carbon sequestration is included in Figure 7.10. It is apparent that if Zimbabwe could capture a market for carbon, then the value of the forest for local use, plus carbon, is similar to that of cropland (Kundhlande et al, 2000).

The modelling indicated that the expulsion of the squatters who are currently in the state forest would have little ecological impact on the forest, largely because their density is relatively low. If this expulsion is combined with the exclusion of local people's cattle, then the ecological system will, if anything, degrade. The expulsion of cattle would result in higher fuel loads and greater frequencies of hot fires, which reduce tree regeneration. Economically, the Zimbabwean Forestry Commission does not increase the value it derives from the forest by expelling the forest dwellers (see Figure 7.11), because expulsion and maintaining the forest free of people would mean increased enforcement costs.

Source: Campbell et al, 2000a

---

hold decision-making is optimized by allowing decisions to vary over time in relation to fluctuating resource availability, changing economic opportunities and risk (Luckert et al, 2000). A third weakness was the inadequate representation of all the relevant disciplines. The sociological–anthropological perspective on resource use was lacking, so that the complexities of local institutional control on resource use, as highlighted in the writings of Mukamuri (1995) and Mandondo (2001), were not reflected. To incorporate this would require considerable effort, both to engage the relevant sociologists and anthropologists, who do

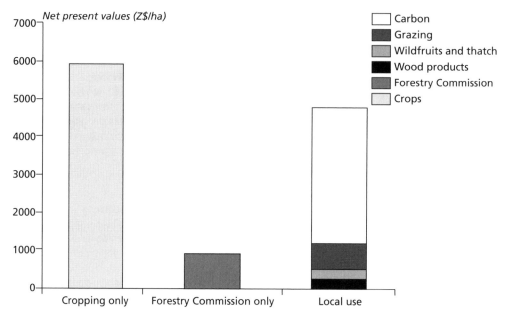

Note: At the time of the study US$1 = Z$10.
Source: Campbell et al, 2000b

Figure 7.10 *Net present values for various land uses, calculated on a per hectare basis for a 60-year period at 6 per cent discount rates. The cropping value is for maize production, the Zimbabwean Forestry Commission value is for grazing leases and timber concessions, and the local-use value is for subsistence use of the forest (grazing, thatch, wild fruits, wood) and carbon sequestration values*

not usually address problems from a modelling perspective, and to translate their knowledge into sets of rules that govern and constrain the processes of resource use.

Models also have other more general limitations. Some models have a large number of parameters that must be specified before they can be run and tested. With complex models, it is often difficult to determine why a particular result arose, and whether it is robust or simply a consequence of some misrepresented or omitted component or process. Some crucial elements may initially be overlooked, particularly if they occur infrequently or episodically. With simple models, or models in the early stage of development, it is usually relatively easy to identify and

incorporate these missing elements; but this becomes harder to do as the models become more complex. Disparities in the depth of knowledge of different components and processes may bias a model or limit its usefulness for predictive purposes, although often it is only through modelling that the significance of any imbalances becomes clear.

Too often, modelling is viewed as an activity to be carried out after all the data have been collected and analysed, or its utility is questioned on the grounds that the model oversimplifies reality. On the contrary, building a model at the outset often helps to ensure that time and resources are focused on collecting relevant data rather than data that might prove to be irrelevant or superfluous to the

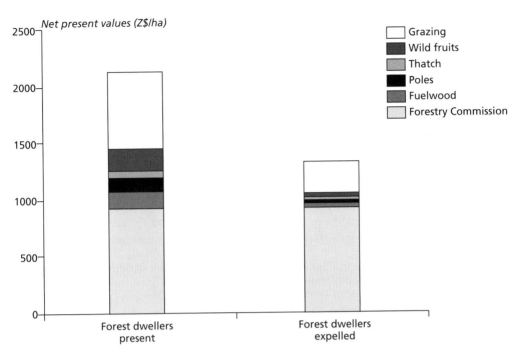

Note: At the time of the study US$1 = Z$10.
Source: Campbell et al, 2000a

Figure 7.11 *Net present values for the state forest when forest dwellers are present and excluded, calculated on a per hectare basis for a 60-year period at 6 per cent discount rates. The Zimbabwean Forestry Commission value is for grazing leases and timber concessions, and the local-use value is for subsistence use of the forest by forest and/or communal dwellers (grazing, thatch, wild fruits, wood)*

eventual enhanced model. By focusing on the few presumed key features and by omitting much of the detail, modelling parallels the selectivity of the senses; perception is not indiscriminate. Nevertheless, the underlying simplifying assumptions should be clearly indicated when reporting the outputs. To be credible, models also need to be validated in order to establish the bounds of their utility; however, even flawed models can be useful if they reveal the limits to our knowledge and understanding.

Modelling is therefore a valuable tool for interdisciplinary integration and synthesis. Given the breadth and complexity of the issues involved in sustainable develop-

ment, there is a need to increase modelling capacity to address the issues in an integrated way and to reverse the prevailing emphasis on reductionism. Nevertheless, while modelling can provide a forum for interactions among disciplines, not everyone is equally comfortable with the process – particularly the need to simplify and generalize. However, with the new modelling software packages that are available, the basics of modelling (and its capacity to promote new ways of conceptualizing and addressing problems) are available to anyone with the interest to try (High Performance Systems, 1993; see Box 7.7).

Box 7.9 Rattan farmers of East Kalimantan: using GIS to understand livelihood strategies

Sonya Dewi

Rattan has been widely used in East Kalimantan as a source of income (see Figure 7.12). Farmers plant rattan seeds or seedlings, along with rice, in their dryland rice fields (*ladang*) and leave the rattan to grow with other shrubs and trees during the fallow period as an important part in the cycle of shifting cultivation.

During recent years, there appears to be a shift away from rattan as a major livelihood source. To understand this pattern a survey was conducted in 53 villages. The villages were classified as being in one of three categories with respect to rattan as a major livelihood source:

1   Villages still using rattan actively, and managing rattan by planting and tendering ('active rattan').
2   Villages where rattan is still an important livelihood source but where there is little active management of the resource ('standby rattan').
3   Villages where rattan is no longer an important livelihood source ('ex-rattan').

Data collected from the villages included numbers of households, major ethnic group, education status, general wealth status, fire impact and the existence of plantations and other projects. Accessibility to markets is difficult to measure quantitatively because of the different kinds of markets that exist (eg large versus small markets; markets for specific products) and the different kinds of transport routes (eg river versus road; poor-quality roads versus good-quality roads; all-weather versus seasonal roads). GIS can be used to rapidly generate variables that reflect accessibility to different kinds of markets, and can be used to explore the spatial patterns of variability among villages.

Spatial dependence is known to be an important aspect in land-use changes and is crucial in this study. The distance of each village, both through river and road transportation, to the closest facility, such as a sub-district capital and town, was calculated using network analysis and shortest network paths' extension in ArcView GIS 3.2.

Statistical analysis using non-linear principal components analysis (PCA) on the socio-economic and spatial variables showed that spatial variables are the most important elements in determining the shift of importance of rattan as the major livelihood source (see Figure 3.7 for an example of this technique). If a village is classified as 'active rattan' (ie actively involved in the use of rattan), then it is highly likely that neighbouring villages will be 'active rattan' as well. The distance by river networks to the nearest towns that are accessible by road, and the distance to the nearest sub-district capitals and towns, by road networks, are two other variables that influence the rattan status of villages. The more remote the villages are from sub-district capitals and towns, the more likely they will remain 'active rattan' villages.

Socio-economic variables such as major ethnic group and education are also strongly correlated with rattan status; however, these variables correlate strongly with the accessibility variables mentioned above. Kutai District tends to have more villages with active rattan than Pasir District, and this factor is also spatially dependent – namely, accessibility to the main settlement in Kutai District is much poorer than that in Pasir District.

The study concludes that with road development, the 'active rattan' villages will change to 'ex-rattan', at least if external variables, such as prices for rattan, remain unchanged. Moreover, given the spatial dependence, a domino effect is likely to take place, with groups of geographically linked villages changing status at similar times. Other sources of income, such as gold mining, fishing and coffee and oil palm production, are likely to become increasingly important.

Photographs by (a) Paul Sochaczewski, (b) Yani Saloh and (c) Brian Belcher

Figure 7.12 *Rattan forms a major source of livelihood in many parts of Indonesia and is an important export for the country. Here we see (a) rattan being dried in Sumatra, (b) rattan being cleaned in East Kalimantan, and (c) one of the final products of rattan, a chair for sale in East Java*

Use and adoption of geographical information systems (GIS)

Geographical information systems have emerged as an integrative tool in research. The different layers of data that are used within a GIS framework allow for the integration of different data derived from the social as well as the physical sciences. GIS can be used at different scales in the valuation of forests. At the scale of a village, homestead sites can be recorded and overlaid on resource maps, so that it is possible to develop an understanding of how resource values differ among households in relation to spatial patterns of people and their resources. At a larger scale, villages can be situated within regional economies in order to understand the patterns of marketing forest products, as well as local livelihood patterns (see Box 7.9).

Constraints to interdisciplinarity

There is much justification for interdisciplinary research, and it can be particularly fruitful when social and physical scientists mix well (Chambers, 1983). However, truly integrated interdisciplinary research is rare. There are many problems – some that are conceptual and others that are methodological. Economists have coined the term transaction costs, which include the costs (mostly time) related to negotiation, discussion and conflict resolution between partners. The transaction costs of interdisciplinary work are exceptionally high. Most disciplines have rich histories that include the development of specialized terminology, concepts and a range of analytical methods. It is unlikely that many people can master the skills and techniques of two or more disciplines. Therefore, to maintain depth of analysis within a discipline, any good interdisciplinary work will normally involve specialists from two or more disciplines who not only work together but who consciously try to bridge the gap between their disciplines. This is in order to see where they interface, and to explore the nature and implications of the interactions across that interface. There are also serious problems related to the specialized language of the disciplines. Words such as system,

regime and niche have been used across disciplines and are interpreted differently. The problem with the word 'institution' has been highlighted in the section on 'Institutions'. An initial step in interdisciplinary work is to develop a common conceptual and methodological framework. This is not simple. Biophysical scientists, for example, when she is bombarded with an array of new terms that may mean something other than their everyday connotations (institutions, governance structures, institutional arrangements, tenure niches, organizations, narratives, value, etc), may tend to retreat into their own discipline. As a further example, social scientists tend to be put off by the efforts of the biophysical scientists to achieve precision and quantitative rigor. Interactions between disciplines are therefore sometimes seen as compromising good science.

The frustration of interacting with other disciplines can be seen in the oft-quoted statement by a biophysical scientist made at a workshop: 'Why do I need to know how many grandmothers per hectare there are before I make a soil fertility recommendation?' The emphasis on 'political correctness' by social scientists also makes interdisciplinary interaction difficult, given that biophysical scientists will often regard 'people' as extraneous to their work, or seek to generalize about them in ways that make social scientists feel uncomfortable.

In one respect, researchers in developing countries have an advantage over their counterparts in developed countries – there is typically less within-discipline competition than is found in Europe or North America, for example. Thus, researchers are freer to cross disciplinary boundaries. This comes about because there are fewer researchers in the region and the pressure to publish is not as great. For instance, a colleague in the UK relates how his job description in his first year of work is to publish three papers in the top ten economics journals. To do that he has to increase the theoretical content of the draft manuscripts and strip them of any non-economic data (eg data collected in participatory rural appraisal). Such a research environment does not foster interdisciplinary work.

Conclusions

The need for an interdisciplinary approach to research is undisputed. Interdisciplinarity is necessary in order to deal with complexity and real world circumstances. Achieving true interdisciplinarity is often difficult and costly. The transaction costs, in particular, can become prohibitive. Although many kinds of research are referred to as being interdisciplinary, most are better described as being multidisciplinary. Real interdisciplinarity focuses on interactions at the interface of disciplines, and on the potential for components and processes in one disciplinary field to significantly affect the functioning of those in others. Successful interdisciplinarity research is still very much the product of a lucky combination of personalities, shared interests, good communication and a desire to make a difference.

Chapter 8

Expanding our conceptual and methodological understanding of the role of trees and forests in rural livelihoods

Martin K Luckert and Bruce M Campbell

Introduction

Discussions in previous chapters highlight the complexity of the methods and underlying tasks in valuing forest resources. In many cases several techniques will be needed to tackle the valuation of specific resources (see Boxes 1.2 and 2.3 for a demonstration of a wide range of methods to derive values in particular case study areas). Even though estimating resource values may seem daunting, if done carefully it can be extremely useful. Indeed, the potential for valuation efforts to inform development initiatives is bound to increase over time, as more is learned regarding when, and how, valuation methods may be applied. Although the application of many of these methods to developing-country settings is in its infancy, the techniques are continually and rapidly improving.

Even as the reliability and precision of value estimates improve, there are other areas of inquiry that may enable better use of value estimates. As has been emphasized many times in this book, estimates of forest values may enable researchers to inform policy-makers, donor agencies and local decision-makers who are attempting to plan and implement interventions that improve livelihoods.[1] Such a process is portrayed in Figure 8.2 and represents, generally, the current state of how information regarding value estimates is used (see Boxes 1.1 and 1.2 for typical uses of valuation figures). However, if inquiries are halted after having informed decision-makers of the estimated resource values, this will have unnecessarily limited the potential role of research to inform policy. In order to use estimates of resource values more fully, it is important to link values of resources with patterns of behaviour. In addition, these patterns of behaviour must be linked with changes in livelihoods.[2]

When organizations design a project or policy, they are generally interested in knowing the effects of the intervention on the welfare of local people (Carney, 1998). However, such projects do not merely hand out money to increase people's welfare, or similarly touch people with magic wands and make them better off, as may be implied in Figure 8.2. Projects expect to change the physical and/or socio-economic environments of people, to allow them to improve their situations.

Photograph by Alain Compost

Figure 8.1 *A Dyak in West Kalimantan returning from a rattan-collecting trip. While in the forest, he will also be on the lookout for opportunities to hunt game, harvest wild plant foods and collect honey. Within the livelihood system of such households, time will also be allocated to crop and tree production. Valuation is only one small step towards understanding the livelihood system of such households*

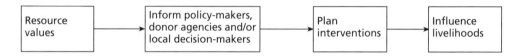

Figure 8.2 *Potential influence of valuation estimates on livelihoods*

Interventions in social environments frequently include changes in public policy and property right arrangements, or the development of local institutions. Interventions in physical environments often include the introduction of new natural resource-management techniques ('technologies'), including technologies that accommodate variable or changing climatic patterns, improve resource-use efficiency or provide higher output levels.

Figure 8.3 adapts Figure 8.2 to add in explicit changes to environments and behaviour. In Figure 8.3, the plans of those attempting to facilitate development are shown to change the environments of people and thereby influence the values of resources. These changes in values influence behaviour, which in turn influences livelihoods. Unlike in Figure 8.2, all influences created by policy-makers and donor agencies are shown to feed back to these decision-makers as they monitor, assess and plan further interventions.

Whether and how improvements in livelihoods occur depends upon what people are able to do within their transformed environments. Accordingly, understanding how people will behave in response to changes in their environment is crucial to assessing whether interventions improve livelihoods. For example, we may chose to introduce a tree-planting programme, but unless we understand something about the behaviour of people, we may introduce a project that has little interest to local people. Alternatively, we may find that a tree-planting project is structured, so that the benefits are only received by select segments of society, which may preclude women and the landless. Accordingly, it is useful to consider behavioural models of the people who are targeted by such projects.

Models of behaviour

Economists studying behaviour assume that people in developing countries tend to do what they do for very good reasons – in other words, that there is carefully tried, tested and chosen logic behind their actions. Accordingly, the actions of rural households may be considered to be rational, as they act in logically consistent ways to pursue their objectives.

The values that people attribute to forest resources are key to understanding this rational behaviour. For example, if a person decides to undertake a craft-making activity rather than planting trees, then craft-making is presumably of more value to the person than tree planting. However, if a development project were to subsidize the tree-planting alternative, so that the value of tree planting exceeds returns to making crafts, then behaviour may change. Whether or not change occurs depends upon a number of factors, other than the development project, which influence resource values and the subse-

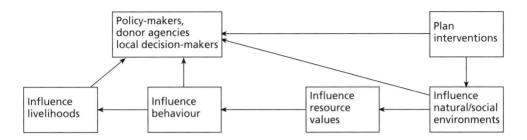

Figure 8.3 *Potential influence of valuation estimates on livelihoods through behaviour*

quent behaviour of local people. These factors include property rights, social differentiation, time, risk perceptions and whether values are considered in 'partial' or 'general' contexts.

Influences of property rights on values and behaviour

Property rights may be thought of as the rules that govern the use and enjoyment of goods and services. These rules generally exist at multiple hierarchical levels, ranging from macro-level national legislation, to local-level village rules. Indeed, there may even be property rights implicit in local norms or courtesies with regard to resource use (see, for example, Sethi and Somanathan, 1996).

Property rights influence the values that people attach to natural resources by influencing what people may and may not do as they enjoy goods and services. Accordingly, values of natural resources are coloured by the property rights that pertain to them. In other words, estimates of resource values are endogenous to the specific property rights that govern the use of the natural resources in question. It follows, then, that changes in property rights can change values of resources. Changing property rights is, therefore, a potential intervention that changes a social environment, influences resource values and alters behaviour and livelihoods (see

Figure 8.3). For example, suppose that a donor agency is considering implementing a tree-planting project. If the agency attempts to promote planting in areas of communal woodlands, it may be difficult to convince people to plant, as local property rights may dictate that trees may be used and/or harvested by others in the village, or even by people from outside villages. Conversely, if tree-planting efforts are targeted towards homestead gardens, where households have exclusive rights, people may be more willing to participate. The general point is that property rights are crucial elements of incentive frameworks for behaviour, and therefore key in influencing whether interventions improve livelihoods.[3]

Influences of social differentiation within households on values and behaviour

Research has shown repeatedly that values of resources, and potential influences of projects, are likely to vary among wealth classes, age classes and between genders (see 'Emerging results' in Chapter 2; see, also, Martin, 1995; Arnold and Ruiz Pérez, 1998; Cavendish, 1997; Clarke et al, 1996; Goebel et al, 2000). Accordingly, it is usually necessary to estimate values for different segments of society.

For some purposes, analysts may choose to estimate values at a household level, where within-household differences in ages and gender are not explicitly considered (see 'The household as the unit of analysis and handling absentee members' in Chapter 2). With this approach the household is treated as a rational entity that seeks to pursue a set of common objectives with a common pool of resources. However, such an approach does not provide information about how different members of a household, such as women, men and children, behave and benefit from interventions. If this type of data is desired, sub-household level estimates must be made.

Influences of time on values and behaviour

Another important variable that influences resource values is time. Values that people attribute to resources do not only arise from quantities and qualities of goods and services: they also arise when a good or service is consumed. All other things being equal, people generally prefer to consume sooner rather than later, causing resource values that pertain to the future to be discounted (see 'Choosing a discount rate' in Chapter 5 for a discussion on discount rates).

Understanding the influence of time on resource values can be critical to understanding behaviour and livelihoods (see, for example, Randall, 1987). Using the tree-planting example again, suppose that a donor agency is trying to convince people to plant trees. The agency has no success, despite the fact that the benefits in ten years' time are valued at $10 per tree, and the input costs required of the household are only $1 per tree. In this case, it may be that the project has no participants because the future is discounted to the extent that the $10 in ten years is not worth as much as the $1 cost incurred now.

Thinking explicitly about values in relation to time also brings up questions regarding changing resource values. Values are known to change over time in response to a host of factors, including changes in preferences and incomes on the demand side, and changes in the costs of production (influenced by changes in availability, technology and the natural environment) on the supply side. Such variability over time can be a mixed blessing for the analyst. Changing values makes predictions regarding future values, behaviour and livelihoods more difficult. However, variability in the past may provide the opportunity to understand why values alter. Given the complexity of the changes, methods such as simulation modelling may become valuable tools (see 'Modelling' in Chapter 7).

Influences of risk perceptions on values and behaviour

People in developing countries usually live in highly variable social and natural environments. Social environments may change due to political instability, rapid technological change or rapid change in the macro-economy (see Figure 1.4). Natural environments change because of cycles of drought or changing availability of resources (see Figure 8.4). In the face of such changes, people can be expected to adopt logical strategies in order to cope with fluctuation and uncertainty.

In developed countries, there are numerous means of diversifying livelihood bases, such as saving funds in banks or mutual funds. However, in many developing countries, such saving opportunities are not available. People often deal with risk through decisions regarding the use of natural resources. For example, people may rely heavily on indigenous trees and forests as safety nets during times of drought (Chambers, 1988; Arnold and

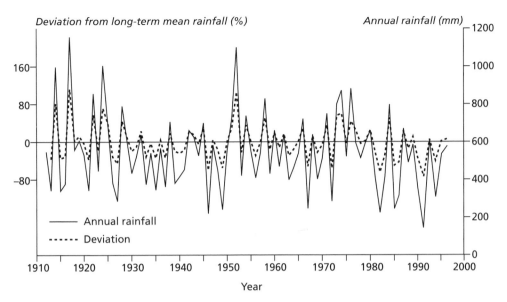

Figure 8.4 *The rainfall in Chivi, Zimbabwe, reflecting the tremendous annual variability. This variability reflects considerable risks for households in that they include livestock and crop production in their portfolio of activities*

Ruiz Pérez, 1998; Wollenberg and Nawir, 1998). Accordingly, if the tree-growing project, mentioned above, were continued, a donor agency may find resistance in a project that converts natural woodlands into plantations; such plantations do not serve as well in terms of ameliorating risks.

The importance of risk in influencing resource values suggests that it is important to consider household livelihood activities as portfolios, where the value of a resource is not assessed in isolation, but is considered as part of a livelihood package. It is this package, rather than individual activities, that discloses peoples' strategies for dealing with fluctuating environments.

Influences of 'partial' or 'general' contexts on values and behaviour

The importance of exploring values of natural resources in the context of multiple activities may be important for reasons other than those discussed in the previous section. Assessing values, behaviours and livelihoods that are associated only with one resource ('partial' analyses) may neglect important interrelationships in situations where values, decisions and livelihoods play a role as part of many potential activities ('general' contexts). A particular behaviour can be thought of as the best of a number of possible choices. Therefore, whether a person is attracted by an opportunity to use or manage a resource will depend upon the values of the resource in question relative to other opportunities. In the tree project example, a donor agency may have determined that the benefits of participating in the tree planting activity are greater than the costs. However, there may, nonetheless, be no participants because the returns are not as high as from other activities. In this case, the partial context assessment would not disclose the reason for non-participation that is evident in the general context.[4]

In addition, values of resources vary according to how much of a resource a person already has. The benefits which individuals receive from a good or service may decrease as a person acquires more of that good or service.[5] In other words, each additional unit of a good received, such as a wild fruit, may be worth less than the previous one. Indeed, this is a key reason why people may undertake a suite of activities rather than specialize in one endeavour. In the absence of diminishing values, it may be more beneficial to produce that one thing (eg the collection of a wild fruit) that maximizes returns. These changes in values of a resource, crucial to understanding behaviour, are frequently not evident in a partial context, and may only become evident in a general context setting.

Trying to address the whole system

Although the focus of this book is on methods of valuing natural resources, the preceding discussion points out that estimating values is not as important as interpreting them within specific contexts.[6] These contexts provide insights into why people do what they do, and whether the changed behaviours created by changing natural and/or social environments improve livelihoods.

There are three principle issues regarding the context of values. Firstly, conceptual models of people and their resources, in terms of livelihoods, need to be expanded (see the following section). Secondly, it is important to consider relevant systems beyond socio-economic variables (see 'Conceptions of relevant systems'). Thirdly, methodological boundaries need to be further developed (see 'Expanding the methodological boundaries').

Conceptions of livelihood

Much of the focus of this book has been on attempts to numerically estimate values of resources, generally in monetary terms (see, for example, Chapters 2 and 4 with their focus on market and non-market valuation, respectively). In order to focus our descriptions on methods, concepts of livelihoods have been necessarily narrow. Indeed, in Figure 8.3 a single box represents the conception of livelihood. However, in recent discussions, 'livelihood' is comprised of five 'capitals' (Bebbington, 1999; Carney, 1998): produced capital, natural capital, social capital, cultural capital and human capital (see Figure 7.8). The focus of this book has been on the methods for generating data on only two of the 'capitals': produced capital and natural capital. Apart from embracing studies that attempt to understand the full spectrum of capital assets, it is important to understand how these assets are influenced by the external biophysical or external context (shocks, trends and seasonality). It is also crucial to understand how these assets translate into livelihood strategies and outcomes, which in turn feed back to the assets (see Figure 8.5). An interdisciplinary perspective is implicit in moving towards this all-embracing conception of livelihood, and involves expertise from several disciplines (see Chapter 7).

Conceptions of relevant systems

The preceding section on 'Influences of "partial" or "general" contexts on values

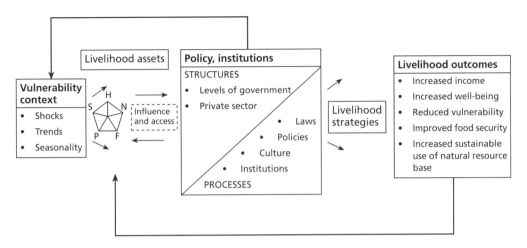

Note: This figure by Carney (1998) also illustrates how Bebbington (1999) and Carney differ in their conception of livelihood. Bebbington (see Figure 7.8) differs by including cultural capital, while lumping financial and physical capital into produced capital.
Source: Carney, 1998

Figure 8.5 *A livelihoods framework (H = human capital; N = natural capital; F = financial capital; P = physical capital; S = social capital)*

and behaviour' has already discussed the potential advantages of considering general rather than partial concepts in the context of values and behaviour. This discussion can be expanded to apply to larger views of systems, rather than just the value/behaviour components. One means of addressing a larger system is to embrace concepts such as integrated natural-resource management (INRM). Although definitions of INRM vary, some of the key components are shown in Figure 8.6. In this figure, it is evident that, in addition to the value/behaviour components discussed above, the relevant context likely includes ecological components. The household or the individual are portrayed as decision-makers who make management decisions in relation to their objectives and the objectives and activities of other stakeholders in the context of ecological, economic and socio-political variables. The management results in a series of outputs and impacts on the environment that may feed back and alter

the context; this subsequently influences future management decisions. Once again, in order to understand such a system, an interdisciplinary perspective is essential (see Chapter 7).

Expanding the methodological boundaries

With our new visions of the INRM system and a broader concept of 'livelihood', we need to incorporate methods beyond those of valuation. Some of these include geographical information systems (GIS), systems modelling and various data management techniques (see the sections on 'Modelling' and 'Use and adoption of geographical information systems (GIS)' in Chapter 7). The importance of considering data in a spatial sense has become increasingly evident. The spatial variability of resources and the impacts on resource values have been shown in Box 4.9. Systems modelling forces researchers to make explicit their concept of the

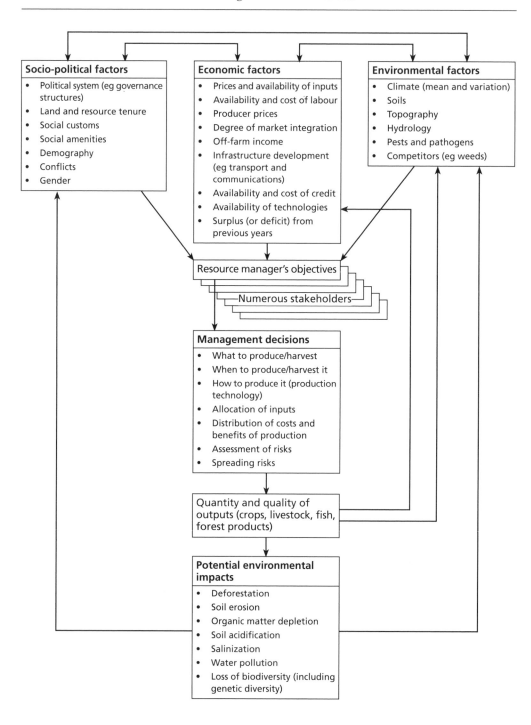

Source: Peter Frost, pers comm, adapted from Swift et al 1994

Figure 8.6 *A conceptual model of an integrated natural resource-management system*

system that they are dealing with, assists in identifying the areas where there are gaps in knowledge, and can provide a means of modelling changes in a complex system (see Boxes 7.7 and 7.8).

However, two words of caution should be added. Firstly, despite the potential for cross-disciplinary modelling, it should be noted that many of the difficulties of performing interdisciplinary work (mentioned in Chapter 7) become evident during modelling exercises. Specifically, some disciplines that seem to rely more heavily on empirics, rather than generalized theory, may have difficulty expressing their concepts in a modelling setting. Secondly, in other cases, the modelling framework is so easy to use that the temptation is to overcomplicate models, exceeding what the knowledge and data can support. With each added component in a system, the potential for interactions among system components increases exponentially. The added complexity and the need for specialists in other disciplines (eg information science) is not without increased costs to researchers; strategic decisions therefore have to be made about what components to include in a system.

A further complication that arises as researchers and disciplines increase is data management. In some cases, it may prove necessary to build an information system that can be a source of data to all participants. With advances in database capabilities and the Internet, such systems are now becoming more central to research projects.

Facilitating development beyond the 'Field of Dreams'

In a popular Hollywood movie, *The Field of Dreams*, the hero (who is considering developing a baseball field) repeatedly hears a voice telling him: 'If you build it, they will come.' A historic view of development efforts displays much of the same philosophy. Unsupported assumptions, or perhaps voices from within, have plagued many projects that have ignored the human dimensions of development. Such projects have been technologically driven by people who are convinced that their particular innovation is just what is needed by local people, with little or no knowledge of local values. Although problems associated with development in this 'Field of Dreams' context have been recognized for some time, progress towards facilitating development beyond the dream has been slow.

Part of the delay in balancing technical knowledge with social knowledge has arisen from a lack of capacity to address resource valuation issues. In a field previously dominated by medical, engineering and natural science technology, expertise dealing with local development issues regarding social values and institutions was sorely lacking. Not only did developing countries lack the capacity to develop social scientists who were versed in development issues, but developed countries were, in many cases, in not much of a position to help. While anthropologists have been studying cultures in developing countries for some time, disciplines such as economics are much newer in attempting to address issues involving natural resource values and household livelihoods in developing countries.[7]

The infancy of using applied economics tools in developing-country settings is evident throughout this book. Indeed, many of the tools are so new and complex

237

that their use is still sometimes hotly debated within developed-country settings. This implies that only a few practitioners, currently at the cutting edge of developing these approaches, are capable of reliably using some valuation methods. Considering this problem within the added complexities of cross-cultural value differences, subsistence economies, local institutions and ecological systems makes the task of resource valuation seem almost intractable.

Nonetheless, the examples presented in this book show that great strides have been made, largely through the benefits of interdisciplinary work. Economics experts, specializing in valuation methods and property rights, have been working with sociologists, anthropologists and ecologists to address the complexities inherent in introducing human dimensions into research in order to inform development projects and policies. We are increasingly becoming better informed about questions such as: 'If we build it, will they come?'

and 'Why did they not come when we built it?' Moreover, an increase in the knowledge and use of participatory methods has changed some of our questions to: 'Can we work together to build something that you would like to visit?'

Although we are still struggling with understanding the social nuances of local peoples, we have nonetheless progressed to the point where social sciences are playing significant roles in balancing technical knowledge in development projects and policies. If we consider that the pursuit of this social knowledge is likely to be a perpetual endeavour, then the importance of maintaining an interdisciplinary approach is heightened, where new types of questions are continuously formed. However, at the same time that we are formulating new questions, we must, if we are to have the financial means of sustaining our pursuit of knowledge, continue to show progress in turning the 'Field of Dreams' into reality.

Notes

1   The concept of improved livelihoods is developed later in this chapter.
2   Angelsen (1999) provides a good critical review of four alternative household models.
3   See, for example, Barzel (1989) and Eggertsson (1990).
4   Note that if all of the relevant opportunity costs are correctly considered in the partial context assessment, the results should coincide with the general context assessment. However, errors in estimated values present in the partial context assessment may only become apparent if there are comparisons that may be made between activities in a general context.
5   This is particularly likely if any extra products made cannot be sold, or if there is significant effort required in selling surplus products.
6   The issue of context was already highlighted in Chapter 6; PRA may be an excellent tool to understand context (see, also, Martin, 1995).
7   Although there is a long-standing field of development economics, this literature deals primarily with macro-economic policy influences on development. Applying micro-economic concepts and tools of natural resource economics to developing country situations is a recent phenomenon.

Acronyms and abbreviations

AIDS	acquired immune deficiency syndrome
BC	benefit-cost (ratio)
Bs	Bolivianos
BMZ	German Federal Ministry of Economic Cooperation and Development
BTU	British thermal units
CAMPFIRE	Communal Areas Management Programme for Indigenous Resources, Zimbabwe
CBA	cost-benefit analysis
C&I	criteria and indicators
CIDA	Canadian International Development Agency
CIFOR	Center for International Forestry Research
Cnd$	Canadian dollar
CV	compensating variation
CV	contingent valuation
CVM	contingent valuation method
DC	district council
DFID	Department for International Development (UK)
DRC	domestic resource cost
ESAP	economic structural adjustment programme
FUG	Forest User Group
EV	equivalent variation
GDP	gross domestic product
GIS	geographical information system
GNP	gross national product
ha	hectare
HP	hedonic pricing
ICE	income, consumption and expenditure
IIED	International Institute for Environment and Development
INRM	integrated natural-resource management
IRR	internal rate of return
ITK	indigenous technical knowledge
kg	kilogram
km	kilometre
m	metre
m^3	cubic metre
MAB	Man and the Biosphere programme
mm	millimetre
n	sample population
N$	Namibian dollar

NDP	national domestic product
NGO	non-governmental organization
NHV	natural habitat value
NPV	net present value
NTFP	non-timber forest product
OECD	Organisation for Economic Co-operation and Development
PCA	principal components analysis
PRA	participatory rural appraisal
R	South African rand
RBG	Royal Botanic Gardens
RDC	rural district council
Rp	Indonesian rupiah
RP	revealed preference
RRA	rapid rural appraisal
SD	standard deviation
SP	stated preference
UK	United Kingdom
UN	United Nations
UNDP	United Nations Development Programme
UNESCO	United Nations Educational, Scientific and Cultural Organization
UNIDO	United Nations Industrial Development Organization
US	United States
US$	American dollar
VIDCO	Zimbabwean village council
WCED	World Commission on Environment and Development
WD	wildlife department
WHO	World Health Organization
WTA	willing to accept
WTAC	willingness to accept compensation
WTP	willingness to pay
WWF	*formerly known as* World Wide Fund For Nature
yr	year
Z$	Zimbabwean dollar
ZVAH	Zambezi Valley Auction Hunts

References

Abbott, J A (1987) *Agricultural Marketing Enterprises for the Developing World*, Cambridge University Press, Cambridge

Abdi F, Hailu, Y and Ayele, K (1997) 'A comparative analysis of participatory approaches: participatory rural appraisal, objectives-oriented project planning, development education and leadership teams in action, local level participatory planning approach and community empowerment programme', SNV-SPADE, Addis Ababa

Acharya, M (1982) 'Time Use Data and the Living Standards Measurement Survey', *LSMS Working Paper* No 18, World Bank, Washington, DC

Adamowicz, W, Luckert, M and Veeman, M (1997) 'Issues in using valuation techniques cross-culturally: three cases in Zimbabwe using contingent valuation, observed behaviour and derived demand techniques', *Commonwealth Forestry Review* 76: 194–7

Adamowicz, W, Beckley, T M, Hatton-MacDonald, D, Just, L, Luckert, M, Murray, E and Phillips W (1998) 'In search of forest resource values of indigenous peoples: the applicability of non-market valuation techniques', *Society and Natural Resources* 11: 51–66

Adams, W D and Megaw, C C (1997) 'Researchers and the rural poor: asking questions in the Third World', *Journal of Geography in Higher Education* 21: 215–29

Ahmed, R and Donovan, C (1993) 'Issues of infrastructural development', International Food Policy Research Institute, Washington, DC

Anderson, A B (1992) 'Land-use strategies for successful extractive economies in Amazonia', *Advances in Economic Botany* 9: 67–77

Angelsen, A (1999) 'Agricultural expansion and deforestation: modeling the impact of population, market forces and property rights', *Journal of Development Economics* 58: 185–218

Arnold, J E M and Ruiz Perez, M (1998) 'The role of non-timber forest products in conservation and development' in E Wollenberg and A Ingles (eds) *Incomes from the Forest: methods for the Development and Conservation of Forest Products for Local Communities*, Centre for International Forestry Research, Bogor, Indonesia, pp17–42

Badiane, O (1999) 'Market integration and the long-run adjustment of local markets to changes in trade and exchange rate regimes', *Agrekon* 38: 353–82

Bain, J (1969) *Industrial Organization*, John Wiley and Sons, New York

Balick, J M and Mendelsohn, R (1992) 'Assessing the economic value of traditional medicines from tropical rain forests', *Conservation Biology* 6: 128–30

Barrett, J C (1992) 'The economic role of cattle in communal farming systems in Zimbabwe', *ODI Pastoral Development Network Paper 32b*, Overseas Development Institute, London

Barzel, Y (1989) *Economic Analysis of Property Rights*, Cambridge University Press, Cambridge

Baumol, W, Panzar, J and Willig, R (1982) *Contestable Markets and the Theory of Industry Structure*, Harcourt, Brace, Jovanovitch, New York

Bebbington, A (1999) 'Capitals and capabilities: a framework for analyzing peasant viability, rural livelihoods and poverty', *World Development* 27: 2021–44

Beckley, T M and Hirsch, B (1997) 'Subsistence and non-industrial forest use in the Lower Liard Valley', *Information Report NOR-X-352*, Northern Forestry Centre, Canadian Forest Service, Edmonton, Alberta

Behnke, R H and Kerven, C (1994) 'Redesigning for risk: tracking and buffering environmental variability in Africa's rangelands', *Natural Resources Perspectives 1*, Overseas Development Institute, London

Behnke, R H, Scoones, I and Kerven, C (eds) (1993) *Range Ecology at Disequilibrium: New Models of Natural Variability and Pastoral Adaptation in African Savannas*, Overseas Development Institute, London

Bishop, J and Scoones, I (1994) 'Beer and baskets: the economics of women's livelihoods in Ngamiland, Botswana', Hidden Harvest Project, International Institute for Environment and Development, London

Blaickie, P (1993) 'What glasses are we wearing? Different views of environmental management' in A Christiansson, A Dahlberg, V Loiske and W Ostberg (eds) *Exploring Interfaces: a Discussion on Natural Resources Research in the Third World*, EDSU, University of Stockholm, pp32–6

Blamey, R, Common, M and Quiggin, J (1995) 'Respondents to contingent valuation surveys: consumers or citizens?' *Australian Journal of Agricultural Economics* 39: 263–88

Blundell, R and Lewbel, A (1991) 'The information content of equivalence scales', *Journal of Econometrics* 50: 49–68

Boadu, F O (1992) 'Contingent valuation for household water in rural Ghana', *Journal of Agricultural Economics* 43: 458–67

Boal, A (1979) *Theatre of the Oppressed*, Pluto Press, London

Bojö, J (1993) 'Economic valuation of indigenous woodlands' in P N Bradley and K McNamara (eds) *Living with trees: policies for forestry management in Zimbabwe*, World Bank Technical Paper No 210, World Bank, Washington, DC, pp227–41

Bond, I (1999) 'CAMPFIRE as a vehicle for sustainable rural development in the semi-arid communal lands of Zimbabwe: incentives for institutional change', DPhil thesis, Faculty of Agriculture, University of Zimbabwe, Harare, Zimbabwe

Bourdillon, M (1987) *The Shona Peoples*, Mambo Press, Gweru

Boxall, P C, Adamowicz, W L, Swait J, Williams, M and Louviere, J (1996) 'A comparison of stated preference methods for economic evaluation', *Ecological Economics* 18: 243–53

Bradley, P N and Dewees, P (1993) 'Indigenous woodlands, agricultural production and household economy in the communal areas' in P Bradley and K McNamara (eds) *Living with Trees: Policies for Forestry Management in Zimbabwe*, World Bank Technical Paper 210, Washington, DC

Braedt, O and Standa-Gunda, W (2000) 'Woodcraft markets in Zimbabwe', *International Tree Crops Journal* 10(4): 367–84

Brigham, T, Chihongo, A and Chidumayo, E (1996) 'Trade in woodland products from the miombo region' in B M Campbell (ed) *The Miombo in Transition: Woodlands and Welfare in Africa*, Centre for International Research, Bogor, pp137–74

Bromley, D W (1980) 'The benefit-cost dilemma', *Western Water Resources*, Papers from a Symposium Sponsored by the Federal Reserve Bank of Kansas City, Westview Press, Boulder, Colorado, pp227–48

Bromley, R J (1980) 'Trader mobility in systems of periodic and daily markets' in D T Herbert and R J Johnston (eds) *Geography and the Urban Environment*, John Wiley and Sons, Chichester, UK

Browder, J O (1992) 'Social and economic constraints on the development of market-orientated extractive reserves in Amazon rain forests' in D C Nepstad and S Schwartzman (eds) *Non-timber Products from Tropical Forests*, Advances in Economic Botany 9, New York Botanical Gardens, New York, pp33–42

Brown, P J and Manfredo, M J (1987) 'Social values defined' in D J Decker and G R Goff (eds) *Valuing Wildlife: Economics and Social Perspectives*, Westview Press, Boulder, Colorado, pp12–23

Brown, W G and Nawas, F (1973) 'Impact of aggregation on the estimation of outdoor recreation demand functions', *American Journal of Agricultural Economics* 55: 246–49

Brown, W G, Singh, A and Castle, E (1964) 'An economic evaluation of the Oregon steelhead fishery', *Technical Bulletin Number 78*, Oregon State University, Agricultural Experiment Station, Corvallis, Oregon

Byron, R N and Bennett, C P A (1998) *Valuing resource valuation: exploring the role of quantitative valuation particularly in developing countries*, Unpublished manuscript. Centre for International Forestry Research, Bogor, Indonesia

Cain, J Moriarty P, Lynam, T and Frost, P (1999) 'An update on the integrating modeling strategy', Micro-catchment management and common property resources, 2nd Integrated Modeling Workshop. Institute of Environmental Sciences, University of Zimbabwe and Institute of Hydrology, UK

Campbell, B M and Mangono, J J (1994) *Working towards a biomass energy strategy for Zimbabwe*, Institute of Environmental Studies, University of Zimbabwe, Harare

Campbell, B M, Costanza, R and van den Belt, M (2000b) 'Land use options in dry tropical woodland ecosystems in Zimbabwe: introduction, overview and synthesis', *Ecological Economics* 33: 341–52

Campbell, B M, Bradley, P N and Carter, S E (1997a) 'Sustainability and peasant farming systems: observations from Zimbabwe', *Agriculture and Human Values* 14: 159–68

Campbell, B M, Doré, D, Luckert, M, Mukamuri, B and Gambiza, J (2000a) 'Economic comparisons of livestock production in communal grazing lands in Zimbabwe', *Ecological Economics* 33: 413–38

Campbell, B M, Frost, P, Goebel, A, Standa-Gunda, W, Mukamari, B and Veeman, M (2000c) 'A conceptual model of woodland use and change in Zimbabwe', *International Tree Crops Journal* 10(4): 347–66

Campbell, B M, Frost, P, Kirchmann, H and Swift, M (1998) 'A survey of soil fertility management in small-scale farming systems in north-eastern Zimbabwe', *Journal of Sustainable Development* 11: 19–39

Campbell, B M, Grundy, I M and Matose, F (1993) 'Trees and woodland resources: the technical practices of small-scale farmers' in P Bradley and K McNamara (eds) *Living with Trees: Policies for Forestry Management in Zimbabwe*, World Bank Technical Paper 210, Washington, DC, pp29–62

Campbell, B M, Luckert, M L and Scoones, I (1997b) 'Local-level valuation of savanna resources: a case study from Zimbabwe', *Economic Botany* 51: 59–77

Campbell, B M, Vermeulen, S J and Lynam, T (1991) *Value of Trees in the Small-scale Farming Sector of Zimbabwe*. International Development Research Centre, Ottawa

Campbell, B M and Byron, N (1996) 'Miombo woodlands and rural livelihoods' in B M Campbell (ed) *The Miombo in Transition: Woodlands and Welfare in Africa*, Centre for International Forestry Research, Bogor, Indonesia

Carney, D (1998) *Sustainable Rural Livelihoods. What Contribution Can We Make?*, Department for International Development, London

Carson, R T (1998) 'Valuation of tropical rainforests: philosophical and practical issues in the use of contingent valuation', *Ecological Economics* 24: 15–29

Carson, R T (1991) 'Constructed markets' in J B Braden and C D Kolstad (eds) *Measuring the Demand for Environmental Quality*, Elsevier, Amsterdam

Carson, R T and Mitchell, R C (1995) 'Sequencing and nesting in contingent valuation surveys', *Journal of Environmental Economics and Management* 28: 155–74

Carter, J (1996) *Recent Approaches to Participatory Forest Resource Assessment*, Overseas Development Institute, Oxford, UK

Cavendish, W (1997) *The economics of natural resource utilisation by communal area farmers of Zimbabwe*, DPhil thesis, University of Oxford, Oxford

Cavendish, W (1999a) 'Empirical regularities in the poverty-environment relationship of African rural households', *Working Paper Series WPS/99.21*, Centre for the Study of African Economies, University of Oxford, Oxford

Cavendish, W (1999b) 'The complexity of the commons: environmental resource demands in rural Zimbabwe', *Working Paper Series WPS/99.8*, Centre for the Study of African Economies, University of Oxford, Oxford

Cavendish, W (1999c) 'Poverty, inequality and environmental resources: quantitative analysis of rural households', *Working Paper Series WPS/99.9*, Centre for the Study of African Economies, University of Oxford, Oxford

Cavendish, W (2001 forthcoming) 'Rural livelihoods and non-timber forest products' in W de Jong and B Campbell (eds) *The Role of Non-timber Forest Products in Socio-Economic Development*, CABI Publishing, Wallingford

Cesario, F J (1976) 'Value of time in recreation benefit studies', *Land Economics* 52: 32–41

Chambers, R (1983) *Rural Development; Putting the Last First*, Longman, London

Chambers, R (1992) 'Rural appraisal: rapid, relaxed and participatory', *IDS Discussion Paper* 311, Institute of Development Studies, University of Sussex, Brighton

Chambers, R and Guijt, I (1995) 'Participatory rural appraisal – five years later. Where are we now?' *Forests, Trees and People Newsletter* 26/27: 4–13

Chambers, R and Conway, G R (1992) 'Sustainable rural livelihoods: practical concepts for the 21st century', *IDS Discussion Paper 296*, Institute of Development Studies, University of Sussex, Brighton

Chambers, R J H (1988) 'Trees as savings and security for the rural poor', *IIED Gatekeeper Series SA3*, International Institute for Environment and Development, London

Child, B A (1988) *The role of wildlife utilisation in the sustainable development of semi-arid rangelands in Zimbabwe*, DPhil thesis, Worcester College, Oxford University, Oxford, p573

Child, B A and Bond, I (1994) 'Marketing hunting and photographic concessions in communal lands' in M A Jones (ed) *Safari Operations in Communal Lands in Matabeleland*, Proceedings of the Natural Resources Management Project Seminar and Workshop, Bulawayo, December 1993. Department of National Parks and Wildlife Management, Harare, Zimbabwe, 37–55

Child, B A (1995) 'A summary of the marketing of trophy quotas in CAMPFIRE areas 1990–1993', CAMPFIRE Coordination Unit, Department of National Parks and Wildlife Management, Harare, Zimbabwe

Clark, J M (1940) 'Towards a concept of workable competition', *American Economic Review* 30: 241–56

Clarke J, Cavendish, W and Coote, C (1996) 'Rural households and miombo woodlands: use, value and management' in B Campbell (ed) *The Miombo in Transition: Woodlands and Welfare in Africa*, Centre for International Research. Bogor, Indonesia, pp101–35

Cliffe, L (1992) 'Towards agrarian transformation in southern Africa' in Seidman, A, Mwanza, L, N Simelane and D Weiner (eds) *Transforming Southern African Agriculture*, Africa World Press, New Jersey, 53–68

Collier, P, Radwan, S and Wangwe, S (1986) *Labour and Poverty in Rural Tanzania: Ujamaa and Rural Development in the United Republic of Tanzania*, Clarendon Press, Oxford

Crawford, I M (1997) *Agriculture and Food Marketing Management*, Food and Agricultural Organization of the United Nations, Rome

Cunningham, A B (1997) 'Review of ethnobotanical literature from eastern and southern Africa', *Bulletin African Ethnobotany Network*, Bulletin 1: 23–87, AETFAT/UNESCO/University of Zimbabwe, Harare, Zimbabwe

Cunningham, A B (2000) *Applied Ethnobotany: People, Wild Plant Use and Resource Management*, Earthscan, London

Cunningham, A B (1990) 'The regional distribution, marketing and economic value of the palm wine trade in the Ingwavuma district, Natal, South Africa', *South African Journal of Botany*

56: 191–98

Dasgupta, A K and Pearce, D W (1972) *Cost-Benefit Analysis*, Macmillan, London

Dasgupta, P (1993) *An Inquiry into Well-Being and Destitution*, Clarendon Press, Oxford

Davies, J and Richards, M (1999) 'The use of economics to assess stakeholder incentives in participatory forest management: a review', *European Union Tropical Forestry Paper 5*, Overseas Development Institute, London

de Beer, J H and McDermott, M J (1996) *Economic Value of Non-Timber Forest Products in Southeast Asia*, Netherlands Committee for IUCN, Amsterdam

de Jong, W, Melnyk, M, Lozano, L A, Rosales, M and Garcia, M (1999) '*Uña de Gato*: fate and future of a Peruvian forest resource', *CIFOR Occasional Paper 22*, Centre for International Forestry Research, Bogor, Indonesia

Deaton, A (1980) 'The measurement of welfare: theory and practical guidelines', *LSMS Working Paper 7*, World Bank, Washington, DC

Deaton, A (1982) 'Inequality and needs: some experimental results for Sri Lanka', *Population and Development Review* 8: 35–49

Deaton, A (1988) 'Quality, quantity, and spatial variation of price', *American Economic Review* 78: 418–30

Deaton, A (1997) *The Analysis of Household Surveys: a Micro-Econometric Approach to Development Policy*, Johns Hopkins University Press, Baltimore

Deaton, A and Paxson, C (1998) 'Economies of scale, household size, and the demand for food', *Journal of Political Economy* 106: 897–930

Diamond, P A and Hausman, J A (1994) 'Contingent valuation: is some number better than no number?', *Journal of Economic Perspectives* 8: 45–64

Dillman, D (1978) *Mail and Telephone Surveys: The Total Design Method*, J Wiley and Sons, New York

Dinwiddy, C and Teal, F (1996) *Principles of Cost-Benefit Analysis for Developing Countries*, Cambridge University Press, Cambridge

Directorate of Forestry (1996) 'Forest biodiversity for present and future generations', Namibia Forestry Strategic Plan, Directorate of Forestry, Ministry of Environment and Tourism, Namibia

Dransfield, J and Manokaran, N (1994) 'Introduction' in *Plant Resources of South-East Asia: Rattans*, Prosea Handbook No 6. Bogor, Indonesia

Eberle, W D and Gregory, H F (1991) 'Critique of contingent valuation and travel cost methods for valuing natural resources and ecosystems', *Journal of Economics Issues* 25: 649–87

Edwards, D M (1996) 'Non-timber forest products from Nepal: aspects of trade in medicinal and aromatic plants', *FORESC Monograph 1/96*, Ministry of Forests and Soil Conservation, Kathmandu

Eggertsson, T (1990) *Economic Behaviour and Institutions*, Cambridge University Press, Cambridge

Ehrlich, P and Daily, G (1993) 'Science and the management of natural resources', *Ecological Applications* 3: 558–60

Emerton, L (1996) 'Valuing the subsistence use of forest products in Oldonyo Orok Forest, Kenya', *Rural Development Forestry Network Paper 19e*, Overseas Development Institute, London

Fletcher, J J, Adamowicz, W L and Graham-Tomasi, T (1990) 'The travel cost model of recreation demand: theoretical and empirical issues', *Leisure Sciences* 12: 119–47

Freeman, A M (1993) *Measurement of Environmental and Resource Values*, Resources for the Future, Washington, DC

Frost, P (1996) 'The ecology of miombo woodlands' in B Campbell (ed) *The Miombo in Transition: Woodlands and Welfare in Africa*, Centre for International Forestry Research, Bogor, Indonesia

Frost, P and Mandondo, A (1999) 'Improving rural livelihoods in semi-arid regions through management of micro-catchments', *IES Working Paper* 12, Institute of Environmental Studies, University of Zimbabwe, Harare

Ganesh, T and Davider, P (1999) 'Fruit biomass and relative abundance of frugivores in a rain forest of southern Western Ghats, India', *Journal of Tropical Ecology* 15: 399–414

Gelfand, M, Mavi, S, Drummond, R B and Ndemera, B (1985) *The Traditional Medical Practitioner in Zimbabwe*, Mambo Press, Gweru

Georgiou, S, Whittington, D, Pearce, D and Moran, D (1997) *Economic Values and the Environment in the Developing World*, Edward Elgar, Cheltenham, UK

Getz, W, Fortmann, L, Cumming, D, du Toit, R, Hilty, J, Martin, R, Murphree, M, Owen-Smith, N, Starfield, A and Westphal, M (1999) 'Sustaining natural and human capital: villagers and scientists', *Science* 283: 1855–56

Godoy, R and Bawa, K S (1993) 'The economic value and sustainable harvest of plants from the tropical forest: assumptions, hypotheses and methods', *Economic Botany* 47: 215–19

Godoy, R and Lubowski, R (1992) 'Guidelines for the economic valuation of non-timber forest products', *Current Anthropology* 33: 423–33

Godoy, R, Brokaw, N and Wilkie, D (1995) 'The effect of income on the extraction of non-timber forest products: model, hypotheses and preliminary findings from the Sumu Indians of Nicaragua', *Human Ecology* 23: 29–52

Godoy, R, Lubowski, R and Markandya, A (1993) 'A method for the economic valuation of non-timber tropical forest products', *Economic Botany* 47: 220–33

Goebel, A (1997) *No spirits control the trees: history, culture and gender in the social forest in a Zimbabwean resettlement area*, DPhil thesis, Department of Sociology, University of Alberta, Edmonton

Goebel, A (1998) 'Process, perception and power: notes from participatory research in a Zimbabwean resettlement area', *Development and Change* 29: 277–305

Goebel, A, Campbell, B M, Mukamuri, B and Veeman, M (2000) 'People, values and woodlands: a field report of emergent themes in interdisciplinary research in Zimbabwe', *Agriculture and Human Values* 17: 385–96

Graham L, Phillips W and Muir-Leresche, K (1997) 'Empirical approaches to the valuation of small wildlife resources in communal areas in Zimbabwe', *Department of Rural Economy Staff Paper 97-04, UA Value of Trees Working Paper #5*, University of Alberta, Edmonton

Green, C (1990) *Canadian Industrial Organization and Policy*, McGraw-Hill Ryerson, Toronto

Grimes et al (1994) 'Valuing the rain forest: the economic value of non-timber forest products in Ecuador', *Ambio* 23: 405–10

Grootaert, C (1982) 'The conceptual basis of measures of household welfare and their implied survey data requirements', *LSMS Working Paper No 19*, World Bank, Washington, DC

Grosh, M E and Muñoz, J (1995) 'A manual for planning and implementing the LSMS survey', *LSMS Working Paper 126*, World Bank, Washington, DC

Grundy, I M, Campbell, B M, Balebereho, S, Cunliffe, R, Tafangenyasha, C, Fergusson, R and Parry, D (1993) 'Availability and use of trees in Mutanda resettlement area, Zimbabwe', *Forest Ecology and Management* 56: 243–66

Gumbo, D J, Mukamuri, B B, Muzondo, M I and Scoones, I C (1990) 'Indigenous and exotic fruit trees: why do people want to grow them?' in R T Prinsley (ed) *Agroforestry for Sustainable Production: Economic Implications*, Commonwealth Science Council, London, pp185–214

Gunawardena, P, Edward-Jones, G, McGregor, M and Abeygunawardena, P (1999) 'A contingent valuation approach for a tropical rainforest: a case study of Sinharaja Rainforest Reserve in Sri Lanka' in C S Roper and A Park (eds) *The Living Forest: Non-market Benefits of Forestry*. Her Majesty's Stationery Office, London, pp275–84

Gunter, J E and Haney, H L (1984) *Essentials of Forestry Investment Analysis*, Oregon State University Press, Corvallis

Gurney, W S C and Nisbet, R M (1998) *Ecological Dynamics*, Oxford University Press, Oxford

Gwaai Working Group (1997) 'Local level valuation of village woodlands and state forests: cases from Matabeleland North in Zimbabwe', *IES Working Paper 7*, Institute of Environmental Studies, University of Zimbabwe, Harare

Haddad, L, Hoddinott, J and Alderman, H (1994) 'Intrahousehold resource allocation: an overview', *Policy Research Working Paper No 1225*, World Bank, Washington, DC

Hanemann, W M (1994) 'Valuing the environment through contingent valuation', *Journal of Economic Perspectives* 8: 19–43

Hanley, N (1992) 'Are there environmental limits to cost-benefit analysis?' *Environmental and Resource Economics* 2: 33–60

Hatton-MacDonald, D (1998) *Valuing fuelwood resources using a site choice model of fuelwood collection*, DPhil thesis, Department of Rural Economy, University of Alberta, Edmonton

Hellerstein, D M (1992) 'The treatment of non-participants in travel cost analysis and other demand models', *Water Resources Research* 28: 1999–2004

High Performance Systems (1993) *STELLA II*, Hanover, New Hampshire

Hill, K A (1993) 'Politicians, farmers and ecologists: commercial wildlife ranching and the politics of land in Zimbabwe', *Journal of African and Asian Studies* 29: 226–47

Hilborn, R and Ludwig, G (1993) 'The limits of applied ecological research', *Ecological Applications* 3: 550–52

Holling, C (1993) 'Investing in research for sustainability', *Ecological Applications* 3: 552–55

Holtzman, J S, Lichte, J A and Tefft, J E (1995) 'Using rapid appraisal to examine coarse grain processing and utilization in Mali' in G Scott (ed) *Prices, Products and People: Analyzing Agricultural Markets in Developing Countries*, Lynne Rienner Publishers, Boulder, Colorado

Hot Springs Working Group (1995) 'Local-level economic valuation of savannah woodland resources: village cases from Zimbabwe', *Hidden Harvest Project Research Series 3*, International Institute for Environment and Development, London

Hot Springs Working Group (2001) *The marketing of products derived by bark in eastern Zimbabwe, IES Working Paper*, Institute of Environmental Studies, University of Zimbabwe, Harare

Howe, C W (1971) *Benefit-Cost Analysis for Water System Planning*, American Geophysical Union, Washington, DC

Hungwe, V D and Fernando, S A S (1996) 'Study on the proposed implementation of a tax on agricultural land in Zimbabwe', A Study for the Ministry of Lands and Water Resources of the Government of Zimbabwe, Harare

Hymam, E L (1996) 'Technology and organisation of production, processing and marketing of non-timber forest products' in M R Perez and J EM Arnold (eds) *Current Issues in Non-Timber Forest Products Research*, Centre for International Forestry Research, Bogor, Indonesia

International Institute for Environment and Development (IIED) (1995) *The Hidden Harvest: The Value of Wild Resources in Agricultural Systems*, IIED, London

IIED (1997) 'Valuing the hidden harvest: methodological approaches for local level economic analysis of wild resources', *Research Series Volume 3, Number 4*, Sustainable Agriculture and Environmental Economics, IIED, London

Jacquemin, A (1987 *The New Industrial Organization: Market Forces and Strategic Behaviour*, The MIT Press, Cambridge, US

Jansen, D J, Bond, I and Child, B A (1992) 'Cattle, wildlife, both or neither? A survey of commercial ranches in the semi-arid regions of Zimbabwe', *WWF (MAPS) Working Paper No 27*, WWF, Harare, Zimbabwe

Jodha, N S (1986) 'Common property resources and rural poor in dry regions of India', *Economic and Political Weekly* 21: 1169–81

Kahneman, D and Knetsch, J L (1992) 'Valuing public goods: the purchase of moral satisfaction', *Journal of Environmental Economics and Management* 2: 57–70

Kengen, S (1997) *Forest Valuation for Decision-Making: Lessons of Experience and Proposals for Improvement*, Food and Agricultural Organization of the United Nations, Rome

Kerstan, B (1995) *Gender-Sensitive Participatory Approaches in Technical Co-operation*, UNICEF, Geneva

Knetsch, J L (1964) 'Economics of including recreation as a purpose of eastern water projects', *Journal of Farm Economics* 39: 387–96

Kramer, R A, Pattanayak, S, Sills, E and Simanjuntak, S (1997) *The Economics of the Siberut and Ruteng Protected Areas: Final Report*, Directorate-General of Forest Protection and Nature Conservation, Government of Indonesia, Biodiversity Conservation Project in Flores and Siberut, Asian Development Bank Loan No 1187-INO

Kramer, R A, Sharma, N, Syamsundar, P and Munasinghe, M (1994) 'Cost and compensation issues in protecting tropical rainforests: case study of Madagascar', *Environment Development Working Paper*, World Bank, Washington, DC

Krutilla, J V (1967) 'Conservation reconsidered', *American Economic Review* 56: 787–96

Krutilla, J W and Fisher, A C (1975) *The Economics of Natural Environments*, Johns Hopkins Press, Baltimore

Kundhlanden, G, Adamovicz, W L and Mapure, I (2000) 'Valuing ecological services in a savanna ecosystem: a case study from Zimbabwe', *Ecological Economics* 33: 401–12

Kwaramba, P K (1995) *Potential Commercialisation of Common Property Resource: The Case of Baobab (Adansonia digitata) Bark around the Hot Springs Area*, University of Zimbabwe, Harare

Lampietti, J A and Dixon, J A (1995) 'To see the forest for the trees: a guide to non-timber forest benefits', *Environment Department Paper 013*, Environment Department, World Bank, Washington, DC

Lanjouw, P and Ravallion, M (1995) 'Poverty and household size', *Economic Journal* 105: 1415–34

Leurs, R (1996) 'Current challenges facing participatory rural appraisal', *Public Administration and Development* 16: 57–72

Loke, S W, Ibrahim, I and Hone, A (1998) 'Report on Market Study of Malaysian Rattan and Bamboo Products' in Lee Su See, D Y May, I D Gauld and J Bishop (eds) *Conservation, Management and Development of Forest Resources, Proceedings of the Malaysia–United Kingdom Programme Workshop, 21–24 October 1996*, Forest Research Institute Malaysia, Kuala Lumpur, 279–89

Long, N and Long, A (1992) *Battlefields of Knowledge: The Interlocking of Theory and Practice in Social Research and Development*, Routledge, London

Luckert, M K and Adamowicz, W (1993) 'Empirical measures of factors affecting social rates of discount', *Environmental and Resource Economics* 3: 1–22

Luckert, M K, Wilson, J, Adamowicz, V and Cunningham, A B (2000) 'Household resource allocations in response to risks and returns in a communal area of western Zimbabwe', *Ecological Economics* 33: 383–94

Luckert, M K, Nemarundwe, N, Gibbs, L, Grundy, I, Hauer, G, Maruzane, D, Shackleton, S and Sithole, J (2001) 'Contribution of baobab production activities to household livelihoods' in Hot Springs Working Group *The marketing of products derived from bark in Eastern Zimbabwe*, IES Working Paper, Institute of Environmental Studies, University of Zimbabwe, Harare

Lynam, T, Campbell, B M and Vermeulen, S (1994) 'Contingent valuation of multipurpose tree resources in the smallholder-farming sector, Zimbabwe', *Working paper Series 1994: 8*, Department of Economics, Gothenburg University, Gothenburg

Mabalauta Working Group (2000) 'The ecology, control and economics of Ilala palm in Sengwe community area, Zimbabwe', *IES Working Paper 15*, Institute of Environmental Studies, University of Zimbabwe, Harare

Mandondo, A (2001) 'Use of woodland resources within and across villages in a Zimbabwean communal area', *Agriculture and Human Values* 18: 177–94

Martin, G J (1995) *Ethnobotany*, Chapman and Hall, London

Matose, F, Mudhara, M and Mushove, P (1997) 'Woodcraft production along the Bulawayo–Victoria Falls Road', *IES Working Paper 2*, Institute of Environmental Studies, University of Zimbabwe, Harare

McConnell, K E (1985) 'The economics of outdoor recreation' in A V Kneese and J L Sweeney (eds) *Handbook of Natural Resource and Energy Economics*, volume II, North Holland Press, New York

McFadden, D (1994 'Contingent valuation and social choice', *American Journal of Agricultural Economics* 76: 689–708

Melnyk, M and Bell, N (1996) 'The direct-use values of tropical moist forest foods: the Huottuja (Piaroa) Amirindians of Venezuela', *Ambio* 25: 468–72

Mendoza, G (1995) 'A Primer on Marketing Channels and Margins' in G Scott (ed) *Prices, Products and People: Analyzing Agricultural Markets in Developing Countries*, Lynne Rienner Publishers, Boulder, Colorado

Messerschmidt, D (1995) 'Rapid appraisal for community forestry', *IIED Participatory Methodology Series*, International Institute for Environment and Development, London

Mitchell, R C and Carson, R T (1984) *A Contingent Valuation Estimate of National Freshwater Benefits*, Technical Report to the US Environmental Protection Agency, Resources for the Future, Washington, DC

Mitchell, R C and Carson, R T (1989) *Using Surveys to Value Public Goods: The Contingent Valuation Method*, Resources for the Future, Washington, DC

Moran, E F (1995) 'Introduction' in E F Moran (ed) *The Comparative Analysis of Human Societies: Toward Common Standards for Data Collection and Reporting*, Lynne Rienner Publishers, Boulder, Colorado

Morris, M L (1995) 'Rapid reconnaissance methods for diagnosis of sub-sector limitations: maize in Paraguay' in G Scott (ed) *Prices, Products and People: Analyzing Agricultural Markets in Developing Countries*, Lynne Rienner Publishers, Boulder, Colorado

Mosse, D (1994) 'Authority, gender and knowledge: theoretical reflections on the practice of participatory rural appraisal', *Development and Change* 25: 497–526

Mudavanhu, H T (1997) *Demography and population dynamics of baobabs (Adansonia digitata) harvested for bark in south-eastern Zimbabwe*, MSc thesis, University of Zimbabwe, Harare

Mukamuri, B (1995) 'Local environmental conservation strategies: Kuranga religion, politics and environmental control', *Environment and History* 1: 63–71

Murphree, M W (1997) 'Strategic considerations for enhancing scholarship at the University of Zimbabwe', *Zambezia* XXIV: 1–11

Murphree, M W (1998) 'Enhancing sustainable use: incentives, politics and science', *Workshop on Environmental Politics*, Working Paper No 99-2, Institute of International Studies, University of California, Berkeley

Murthi, M (1994) 'Engel equivalence scales in Sri Lanka: exactness, specification, measurement error' in R Blundell, I Preston and I Walker (eds) *The Measurement of Household Welfare*. Cambridge University Press, Cambridge, 164–91

Navrud, S and Mungatana, E D (1994) 'Environmental valuation in developing countries: the recreational value of wildlife viewing', *Ecological Economics* 11: 135–51

Ndoye, O, Ruiz Pérez, M and Eyebe, A (1998) 'The markets for non-timber forest products in the humid forest zone of Cameroon', *Network Paper 22c*, Rural Development Forestry Network, London

Ngugi, D, Mataya, C and Ng'ong'ola, D (1997) 'The implication of maize market liberalisation for market efficiency and agricultural policy in Kenya', Paper Prepared for the International Association of Agricultural Economists Conference, Sacramento, California, August, mimeo

North, D C (1990) *Institutions, Institutional Change and Economic Performance*, Cambridge University Press, Cambridge

North, J H and Griffin, C C (1993) 'Water source as a housing characteristic: hedonic property valuation and willingness to pay for water', *Water Resources Research* 29: 1923–29

Ostrom, E (1990) *Governing the Commons: the Evolution of Institutions for Collective Action*, Cambridge University Press, Cambridge

Padoch, C (1992) 'Marketing of non-timber forest products in western Amazonia: general observations and research priorities' in D C Nepstad and S Schwartzman (eds) *Non-Timber Products from Tropical Forests*, Advances in Economic Botany 9, New York Botanical Gardens, New York. pp43–50

Peters, C M, Gentry, A H and Mendelsohn, R O (1989) 'Valuation of an Amazonian rainforest', *Nature* 339: 655–56

Pimbert, M P and Pretty, J N (1995) 'Parks, people and professionals, putting "participation" into protected area management', *Discussion Paper 57*, United Nations Institute for Social Development, Geneva, Switzerland

Pinedo-Vasquez, M, Zarin, D and Jipp, P (1992) 'Community forest and lake reserves in the Peruvian Amazon: a local alternative for the sustainable use of tropical forests', *Advances in Economic Botany* 9: 79–86

Pinedo-Vasquez, M, Zarin, D, Jipp, P and Chota-Inuma, J (1990) 'Use values of tree species in a communal forest reserve in northeast Peru', *Conservation Biology* 4: 405–16

Porter, G (1995) 'Field methods for exploring the role of indigenous rural periodic markets in developing countries' in G Scott (ed) *Prices, Products and People: Analyzing Agricultural Markets in Developing Countries*, Lynne Rienner Publishers, Boulder, Colorado

Powell, N (1998) *Co-management in non-equilibrium systems: cases from Namibian Rangelands*, Swedish University of Agricultural Sciences, Uppsala

Prabhu, R, Colfer, C J P and Dudley, R G (1999) 'Guidelines for Developing, Testing and Selecting Criteria and Indicators for Sustainable Forest Management', *Criteria and Indicators Toolbox Series No 1*, Centre for International Forestry Research, Bogor, Indonesia

Pretty, J, Guijt, I, Thompson, J and Scoones, I (1995) *Participatory Learning and Action, A Trainer's Guide*, IIED Participatory Methodology Series, International Institute for Environment and Development, London

Price, L and Campbell, B M (1998) 'Household tree holdings: a case study in Mutoko communal area, Zimbabwe', *Agroforestry Systems* 39: 205–10

Randall, A (1987) *Resource Economics: An Economic Approach to Natural Resource and Environmental Policy*, John Wiley and Son, Toronto

Randall, A, Ives, B C and Eastman, C (1974) 'Bidding games for valuation of aesthetic environmental improvements', *Journal of Environmental Economics and Management* 1: 132–49

Ravallion, M (1986) 'Testing market integration', *American Journal of Agricultural Economics* 68: 102–9

Richards, M, Davies, J and Cavendish, W (1999a) 'Can PRA methods be used to collect economic data? A non-timber forest product case study from Zimbabwe', *PLA Notes* 36: 34–40

Richards, M, Kanel, K, Maharjan, M and Davies, J (1999b) *Towards Participatory Economic Analysis by Forest User Groups in Nepal*, Report Prepared for Forestry Research Programme, Department for International Development, Overseas Development Institute, London

Riley, H M and Weber, M T (1983) 'Marketing in developing countries' in P Farris (ed) *The Future Frontiers in Agricultural Marketing Research*, Iowa State University Press, Ames

Rohadi, D, Maryani, R, Belcher, B, Ruiz Pérez, M and Widnyana, M (2000) 'Can sandalwood in East Nusa Tenggara survive? Lessons from the policy impact on resource sustainability', *Sandalwood Newsletter* 10: 3–6

Rokeach, M (1973) *The Nature of Human Values*, Free Press, New York

Rowe, R D, Schultze, W D and Breffle, W S (1996) 'A test for payment card biases', *Journal of Environmental Economics and Management* 31: 178–85

Ruiz Pérez, M, Maogong, Z, Belcher, B, Chen, X, Maoyi, F and Jinzhong, X (1999) 'The role of bamboo plantations in rural development: the case of Anji County, Zhejiang, China', *World Development* 27: 101–14

Ruiz Pérez, M and Byron, N (1999) 'A methodology to analyse divergent case studies of non-timber forest products and their development potential', *Forest Science* 45: 1–14

Scarborough, V and Kydd, J (1992) *Economic Analysis of Agricultural Markets: A Manual*, Natural Resources Institute, Chatham, UK

Schaffer, J D, Weber, M, Riley, H and Staatz, J (1987) 'Influencing the Design of Marketing Systems to Promote Development in Third World Countries', *International Development Papers Reprint No 2*, Department of Agricultural Economics, Michigan State University, Michigan

Scherer, F M and Ross, D (1990) *Industrial Market Structure and Economic Performance*, Houghton, Mifflin Company, Boston

Schultz S, Pinazzo, J and Cifuentes, M (1998) 'Opportunities and limitations of contingent valuation surveys to determine national park entrance fees: evidence from Costa Rica', *Environment and Development Economics* 3: 131–49

Scoones, I (1998) *Sustainable Rural Livelihoods: A Framework for Analysis*, Institute of Development Studies, University of Sussex, UK

Scoones, I (1992) 'The economic value of livestock in the communal areas of southern Zimbabwe', *Agricultural Systems* 39: 339–59

Scoones, I and Wilson, K (1989) 'Households, lineage groups and ecological dynamics: issues for livestock research and development in Zimbabwe's communal lands' in B Cousins (ed) *People, Land and Livestock*, Centre for Applied Social Studies, University of Zimbabwe, Harare, pp17–123

Scoones, I and Thompson, J (1994) *Beyond Farmer First: Rural People's Knowledge, Agricultural Research and Extension*, Intermediate Technology Group, London

Scoones, I, Chibudu, C, Chikura, S, Jeranyama, P, Machaka, W, Machanja, B, Mavedzenge, B, Mombeshora, M, Mudhara, M, Mudziwo, C, Murimbarimba, F and Zirereza, B (1996) *Hazards and Opportunities: Farming Livelihoods in Dryland Africa: Lessons from Zimbabwe*, ZED books, London

Scott, G J (ed) (1995) *Prices, Products and People: Analyzing Agricultural Markets in Developing Countries*, Lynne Rienner Publishers, Boulder, Colorado

Serageldin, I and Steer, A (1995) 'Valuing the Environment', *Environmentally Sustainable Development Series No 2*, World Bank, Washington, DC

Sethi, R and Somanathan, E (1996) 'The evolution of social norms in common property resource use', *American Economic Review* 86: 766–88

Shackleton, C M and Shackleton, S E (2000) 'Direct use values of secondary resources harvested from communal savannas in the Bushbuckridge lowveld, South Africa', *Journal of Tropical Forest Products* 6: 21–40

Shackleton, C M, Netshiluvhi, T R, Shackleton, S E, Geach, B S, Ballance, A and Fairbanks, D F K (1999a) *Direct Use Values of Woodland Resources from Three Rural Villages*, Unpublished Report No ENV-P-I 98120, CSIR, Pretoria

Shackleton, C M, Shackleton, S E, Netshiluvhi, T R, Mathabela, F R and Phiri, C (1999b) *The Direct Use Value of Goods and Services Attributed to Cattle and Goats in the Sand River Catchment, Bushbuckridge*, Unpublished Report No ENV-P-C 99003, CSIR, Pretoria

Shanley, P (1999) *As the forest falls: the changing use, ecology and value of non-timber forest resources for Caboclo communities in eastern Amazonia*, DPhil thesis, Durrell Institute of Conservation and Ecology, University of Kent, Canterbury, UK

Shanley, P, Galvao, J and Luz, L (1997) 'Limits and strengths of local participation: a case study in eastern Amazonia', *PLA Notes 28*: 64–7, International Institute for Environment and Development, London

Shapiro, B T and Staal, S J (1995) 'The Policy Analysis Matrix Applied to Agricultural Commodity Markets' in G Scott (ed) *Prices, Products and People: Analyzing Agricultural Markets in Developing Countries*, Lynne Rienner Publishers, Boulder, Colorado

Sharp, C (1993) 'Sooty baobabs in Zimbabwe', *Hartebeest 25*: 7–14

Skinner, G W (1964) 'Marketing and social structure in rural China. Part 1', *Journal of Asian Studies 24*: 3–43

Skinner, G W (1965) 'Marketing and social structure in rural China. Part 2', *Journal of Asian Studies 24*: 195–228

Smith, V K (1988) 'Selection and recreation demand', *American Journal of Agricultural Economics 70*: 29–36

Smith, V K (1989) 'Taking stock of progress with travel cost recreation demand methods: theory and implementation', *Marine Resource Economics 6*: 279–310

Southgate, D D (1998) *Tropical Forest Conservation: An Economic Assessment of the Alternatives in Latin America*, Oxford University Press, New York

Southgate, D, Coles-Ritchie, M and Salazar-Canelos, P (1996) 'Can tropical forests be saved by harvesting non-timber products? A case study for Ecuador', in W L Adamowicz, P C Boxall, M K Luckert, W E Phillips and W A White (eds) *Forestry, Economics and the Environment.* CAB International, Wallingford, 68–80

Staal, S J (1996) 'Public policy and incentives to periurban dairying in Addis Ababa, Ethiopia', Poster presented at XXIII Conference of IAAE, International Livestock Research Institute, Nairobi, Kenya

Stocking, M (1993) 'The rapid appraisal of physical properties affecting land degradation' in A Christiansson, A Dahlberg, V Loiske and W Ostberg (eds) *Exploring Interfaces: a Discussion on Natural Resources Research in the Third World*, EDSU, University of Stockholm, 24–6

Strauss, J and Thomas, D (1995) 'Human resources: empirical modelling of household and family decisions' in J Behrman and T N Srinivasan (eds) *Handbook of Development Economics, Volume IIIA*, Elsevier Science, Amsterdam, 1885–2023

Swift, M J, Bohren, L, Carter, S E, Izac, A M and Woomer, P L (1994) 'Biological management of tropical soils: integrating process research and farm practice' in P L Woomer and M J Swift (eds) *The Biological Management of Tropical Soil Fertility*. John Wiley and Sons, Chichester, UK, 209–227

Tobias, T and Kay, J (1993) 'The bush harvest in Pinehouse Saskatchewan, Canada', *Arctic* 47: 207–21

Trager, L (1995) 'Minimum data sets in the study of exchange and distribution' in E F Moran (ed) *The Comparative Analysis of Human Societies: Toward Common Standards for Data Collection and Reporting*, Lynne Rienner Publishers, Boulder, Colorado

Tschirley, D S (1995) 'Using microcomputer spreadsheets for spatial and temporal price analysis: an application to rice and maize in Ecuador' in G Scott (ed) *Prices, Products and People: Analyzing Agricultural Markets in Developing Countries*, Lynne Rienner Publishers, Boulder, Colorado

Tuxill J and Nabhan, G P (2001) *People, Plants and Protected Areas: A Guide to In Situ Management*, Earthscan, London

United Nations Development Programme (UNDP) (1995) *Human Development Report 1995*, Oxford University Press, Oxford

United Nations Industrial Development Organization (UNIDO) (1972) *Guidelines for Project Evaluation*, United Nations, New York

Vallejos C, Cuéllar R, Ayala J and Ramos, C (1996) *Estudio de valuación del bosque de Lomerío: Memoria del Taller realizado con representantes de las comunidades Las Trancas y Puesto Nuevo*, BOLFOR, Proyecto de Manejo Forestal Sostenible, Ministerio de Desarrollo Sostenible y Medio Ambiente, Santa Cruz, Bolivia

Veeman, M, Cocks, M, Muwonge, A, Choge, S and Campbell, B (2001) 'Markets for three bark products in Zimbabwe: a case study of markets for bark of *Adansonia digitata, Berchemia discolor* and *Warburgia salutaris*' in Hot Springs Working Group *The marketing of products derived from bark in Eastern Zimbabwe, IES Working Paper*, Institute of Environmental Studies, University of Zimbabwe, Harare

Veeman, T S, Cunningham, A B, Kozanayi, W and Maingi, X (2001) 'Muranga returns: the economics of production of a rare medicinal species re-introduced in south-eastern Zimbabwe' in B M Campbell and I Grundy (eds) *The Marketing of Products Derived from Bark in Eastern Zimbabwe, IES Working Paper*, Institute of Environmental Studies, University of Zimbabwe, Harare

Veeman, T S (1989) 'Sustainable development: Its economic meaning and policy implications', *Canadian Journal of Agricultural Economics* 37: 875–86

Veeman, T S (1978) 'Benefit-cost analysis relating to water storage with special reference to the Oldman River Basin in southern Alberta', *Canadian Water Resources Journal* 3: 41–9

Von Braun, J and Puetz, D (1993) *Data Needs for Food Policy in Developing Countries: New Directions for Household Surveys*, International Food Policy Research Institute, Washington, DC

Watson, L Murray, E C and Just, L (1996) 'Gender and the value of trees in Mutoko communal area, Zimbabwe', *Rural Economy Staff Paper 96-20*, University of Alberta, Edmonton

Webb, J L A (1985) 'The trade in gum arabic', *Journal of African History* 26: 149–68

Whittington, D (1998) 'Administering contingent valuation surveys in developing countries', *World Development* 26: 21–30

Whittington, D, Lauria, D T, Wright, A M, Choe, K, Hughes, J A and Swarna, V (1993) 'Household demand for improved sanitation services in Kumasi, Ghana: a contingent valuation study', *Water Resources Research* 29: 1539–60

Whittington, D, Briscoe, J, Mu, X and Barron, W (1990) 'Estimating the willingness to pay for water services in developing countries: a case of the use of contingent valuation surveys in southern Haiti', *Economic Development and Cultural Change* 38: 293–311

Whittington, D, Smith, V K, Okorafor, A, Okore, A, Liu, J L and McPhail, A (1992) 'Giving respondents time to think in contingent valuation studies: a developing country application', *Journal of Environmental Economics and Management* 22: 205–25

Willig, R D (1976) 'Consumers' surplus without apology', *American Economic Review* 66: 589–97

Wollenberg, E and Nawir, A S (1998) 'Estimating the incomes of people who depend on forests' in E Wollenberg and A Ingles (ed) *Incomes from the Forest; Methods for the Development and Conservation of Forest Products for Local Communities*, Centre for International Forestry Research, Bogor, Indonesia, pp17–42

World Bank (1999) *World Development Report: Knowledge for Development, 1998/99*, Oxford University Press, New York

World Commission on Environment and Development (WCED) (1987) *Our Common Future*, Oxford University Press, Oxford

Index

Page numbers in *italics* refere to boxes, figures and tables

Other titles from Earthscan

AGRICULTURAL EXPANSION AND TROPICAL DEFORESTATION
Poverty, International Trade and Land Use
Solon L Barraclough and Krishna B Ghimire.
A multidisciplinary analysis of economic and agricultural development and their impact on increasing land use pressure and change.
Pb £14.95 1 8383 665 6
Hb £40.00 1 85383 666 4

BAD HARVEST
The Timber Trade and the Degradation of the World's Forests
Nigel Dudley, Jean-Paul Jeanrenaud and Francis Sullivan
An incisive account of the role that the timber trade has played in the loss and degradation of forests around the world.
Pb £14.95 1 85383 188 3

BIODIVERSITY AND ECOLOGICAL ECONOMICS
Participation, Values and Resource Management
Luca Tacconi, Australian Agency for International Development
A model of applied ecological economics. It presents an accessible introduction to the subject while applying the theoretical framework to case studies from a range of different ecosystems.
Pb £14.95 1 85383 676 1
Hb £40.00 1 85383 675 3

THE COMMERCIAL USE OF BIODIVERSITY
Access to Genetic Resources and Benefit Sharing
Kerry ten Kate and Sarah A Laird
'This book is an invaluable resource and guide... It is as practical as it is informative and should be in the hands of every policy-maker, entrepreneur and student' Calestous Juma, former Executive Secretary, Convention on Biological Diversity
Hb £50.00 1 85383 334 7

ECOLOGICAL ECONOMICS
A Political Economics Approach to Environment and Development
Peter Soderbaum
'This excellent book is strongly recommended'
Water Resources Development
Pb £15.95 1 85383 685 0
Hb £40.00 1 85383 686 9

ECONOMIC INSTRUMENTS FOR ENVIRONMENTAL MANAGEMENT
A Worldwide Compendium of Case Studies
Edited by Jennifer Rietbergen-McCracken and Hussein Abaza
Describes the diversity of environmental problems to which a variety of economic instruments can be applied, including studies on air and water pollution, packaging, deforestation, overgrazing and wildlife management.
Pb £19.95 1 85383 690 7

THE ECONOMICS OF THE TROPICAL TIMBER TRADE
Edward B Barbier and others
'I strongly recommend this book to all interested in forestry or economic development in tropical areas. It is a superb source of information and should be on all the shelves of all university libraries' *Science Technology & Development*
Pb £14.95 1 85383 219 7

ENVIRONMENTAL VALUATION
A Worldwide Compendium of Case Studies
Edited by Jennifer Rietbergen-McCracken and Hussein Abaza
For non-economists as well as economists, it provides valuable source material for students and academics and for policy makers and professionals using valuation methods to frame policy.
Pb £19.95 1 85383 695 8

THE FOREST CERTIFICATION HANDBOOK
Christopher Upton and Stephen Bass
Practical advice on developing, selecting and
operating a certification programme which
provides both market security and raises
standards of forestry management.
Pb £22.95 1 85383 222 7

FOREST POLITICS
The Evolution of International Cooperation
David Humphreys
Traces the emergence of deforestation as an
issue on the international political agenda. It
assesses the causes of deforestation and its
environmental and social effects, and considers
the problems facing the international community in dealing with the array of issues involved.
Pb £15.95 1 85383 378 9

THE SUSTAINABLE FORESTRY HANDBOOK
*Sophie Higman, Neil Judd, Steve Bass and
James Mayers*
'A stimulating guide to many of the issues likely
to be faced in contemporary forest manage-
ment' James Sowerby, Shell Forestry Ltd
Pb £25.00 1 85383 599 4

TREES, PEOPLE AND POWER
**Social Dimensions of Deforestation in Central
America**
Peter Utting
'An important book which ought to be read by
anyone with a desire to understand the tangled
web of modern environmental concerns'
Habitat
Pb £14.95 1 85383 162 X

TROPICAL DEFORESTATION
A Socio-Economic Approach
C J Jepma
'This fine book will be of interest to natural
and social scientists and resource mangers
interested in the social and environmental
dimensions of the development of tropical
forests... and to environmental economists
interested in the interplay between ecological
systems and human institutions'
Environmental Politics
Pb £18.95 1 85383 238 3

People and Plants Conservation Series

APPLIED ETHNOBOTANY
People, Wild Plant Use and Conservation
Anthony B Cunningham
Pb £24.95 ISBN 1 85383 697 4

BIODIVERSITY AND TRADITIONAL
KNOWLEDGE
Equitable Partnerships in Practice
Edited by Sarah A Laird
Pb £24.95 ISBN 1 85383 698 2
Hb £60.00 ISBN 1 85383 914 0

PEOPLE, PLANTS AND PROTECTED AREAS
A Guide to *In Situ* Management (reissue)
John Tuxill and Gary Paul Nabhan
Pb £24.95 ISBN 1 85383 782 2

PLANT INVADERS
The Threat to Natural Ecosystems (reissue)
Quentin C B Cronk and Janice L Fuller
Pb £24.95 ISBN 1 85383 781 4

TAPPING THE GREEN MARKET
**Management and Certification of Non-Timber
Forest Products**
*Edited by Patricia Shanley, Alan R Pierce,
Sarah A Laird and Abraham Guillén*
Pb £24.95 ISBN 1 85383 810 1
Hb £60.00 ISBN 1 85383 871 3

EARTHSCAN

Tel: +44 (0)1903 828 800   Fax: +44 (0)20 7278 1142   Email: earthinfo@earthscan.co.uk
www.earthscan.co.uk